Practical Protein Bioinformatics

Florencio Pazos • Mónica Chagoyen

Practical Protein Bioinformatics

Florencio Pazos
National Centre for Biotechnology
(CNB-CSIC)
Madrid
Spain

Mónica Chagoyen
National Centre for Biotechnology
(CNB-CSIC)
Madrid
Spain

ISBN 978-3-319-38184-8 ISBN 978-3-319-12727-9 (eBook)
DOI 10.1007/978-3-319-12727-9

Springer Cham Heidelberg New York Dordrecht London

Softcover reprint of the hardcover 1st edition 2015

Printed on acid-free paper

Springer is part of Springer Science+Business Media (www.springer.com)

Introduction

Bioinformatics methods are becoming part of the standard toolboxes of Life Science laboratories. Wisely applied, these approaches can enormously restrict the experimental work and complement the results obtained by "wet" methods. While many bioinformatics methods and protocols are not mature enough to be used by non-bioinformaticians (in terms of implementation, or easiness of the interpretation of the results), many others are already at a stage in which they can be used by non experts. They can be accessed through standard web interfaces from any computer irrespective of its hardware or operative system, they do not require the installation/maintenance of specific software, and their results are easy to interpret and generally presented in a graphical interactive way. In spite of this, our experience shows that many wet labs are not aware of many of the tools freely available.

This book covers these tools that work with information related to proteins. Only the tools fulfilling the type of requirements mentioned above are included. For example, tools requiring the installation of software, or the pre/post processing or parsing of the input or the results are discarded. While the book tries to be exhaustive in covering all aspects related to proteins (from genomic sequences to protein networks, going thorough protein sequences and three-dimensional structures), the number of tools commented in each category is necessarily restricted, and many tools with similar goals are not commented. This selection, personal and biased by definition, is based on our own experience working in a Protein Analysis Facility of a Molecular Biology research centre.

The point of view of the book is totally practical. This is not intended to be an introductory textbook on Bioinformatics. The theoretical bases of the tools are only tangentially mentioned as long as such knowledge is required to better interpret the results. We try to explain the usage of the tools in a similar way protocols are described in Molecular Biology books. Practical examples with real proteins are included in the different sections.

Since the field is moving very fast and many tools get improved, are surpassed, or the web addresses change or simply disappear, the book is associated to a dynamic web site that will try to reflect these changes. This site will maintain an updated list of the tools and include information on their eventual upgrades and changes. This site can be accessed at http://csbg.cnb.csic.es/PB/.

For each tool discussed, a table with the following information is included:

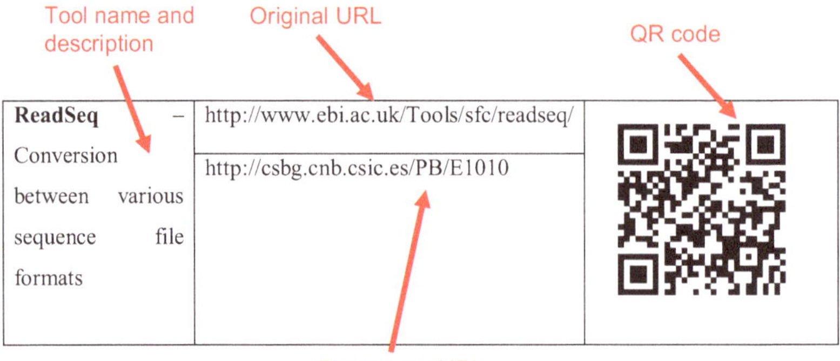

The table contains the name and a short description of the tool, as well as its current web address (URL). A "permanent" URL is also included. Right now that is just an automatic redirection to the original URL of the tool. But in case that original URL changes or disappears, the permanent URL will reflect that, providing information on the change, proposing alternative tools, etc. Finally, the table contains a "quick response" (QR) code that can be scanned so as to automatically point the device browser to the permanent URL. The bibliographic references of the tools are also included so that interested users can obtain more information.

An index with the tools, as well as "How to…?" index have also been included to facilitate localizing the procedure/tool of interest.

Finally, we would like to acknowledge the developers of these tools, who are investing their time and resources for creating and maintaining a large ecosystem of interconnected web applications that facilitates the daily work of molecular biologists.

Contents

Chapter 1
Sequences

1.1 Introduction

An amino acid sequence represents the protein's biochemical composition as a linear polymer built from the covalent attachment of a series of amino acids by means of peptide bonds. It is also referred to as the primary structure of a protein.

Amino acid sequences reflect the exact and unique composition of nascent proteins as they are translated from their mRNA templates. As such they can be thought as the perfect fingerprint to identify a protein, and a valuable source of information that can be used to further infer structural and functional information. They are also the natural link between a protein and its genetic information. In this chapter we present those bioinformatics analyses that deal with protein sequences, from the representation of protein sequences in the computer, to the browsing of the main sequence collections, and the comparison of protein sequences in their different forms: pair-wise and multiple sequence alignments, database searches, and basic phylogenetic analysis.

1.2 Representing Protein Sequences in the Computer

Protein sequences are represented in the computer as a string of characters, using the one-letter notation for amino acid residues established by the IUPAC-IUB (IUPAC-IUB Commission on Biochemical Nomenclature. A one-letter notation for amino acid sequences. Tentative rules. 1969). This notation assigns a letter to each of the 20 natural amino acids, as well as additional characters for representing other features of the sequence (e.g. "X" for residue of unknown nature). The polypeptide chain is always represented from the N-terminus (left, or first character in the string) to the C-terminus (right, or last character in the string).

F. Pazos, M. Chagoyen, *Practical Protein Bioinformatics*,
DOI 10.1007/978-3-319-12727-9_1

Example: Human SPINK1 gene coding for a pancreatic secretory trypsin inhibitor:

```
MKVTGIFLLSALALLSLSGNTGADSLGREAKCYNELNGCTKIYDPVCGTDGNTYPNECVLCFENRKRQTSILIQKSGPC
|                                                                              |
N-term                                                                    C-term
```

1.2.1 *Sequence File Formats*

The file formats most commonly used for storing protein sequences are plain text (ASCII) files. You can open these files in a simple text editor (like Notepad in Windows, or TextEdit in Mac) to read, edit, copy/paste to web forms, etc. Although it is tempting to store and share sequences with colleagues in Word files (.doc or .docx) or PDF files (.pdf), we recommend you not to do so. The reason is that bioinformatics software, both stand-alone and web-based applications, are not able to read these formats (so you won't be able to open or upload these files as input to these programs). It is, therefore more convenient to store and share sequences in plain text files and, always in addition -not as a replacement-, use other type of programs for creating "visual" add-ons or adding manual annotations (like coloring, etc.).

FASTA Format

It is a very simple format that can contain one or multiple sequences. Each sequence is represented with a header line (starting with ">" followed by a string of characters, commonly used to include an identifier and a short description), and the characters representing the sequence in the following lines (up to a new header line or the end of the file). There is no standard file extension, although ".fasta", ".fas" and ".fa" are widely used. Note that FASTA files can contain unaligned or aligned sequences (see Sect. 1.7.1). Take a look at http://en.wikipedia.org/wiki/FASTA_format for an exhaustive description of the format.

Example of a single sequence FASTA file:

```
>sp|P00995|ISK1_HUMAN Pancreatic secretory trypsin inhibitor
MKVTGIFLLSALALLSLSGNTGADSLGREAKCYNELNGCTKIYDPVCGTDGNTYPNECVL
CFENRKRQTSILIQKSGPC
```

Example of a FASTA file with many sequences

```
>sp|P00995|ISK1_HUMAN Pancreatic secretory trypsin inhibitor
MKVTGIFLLSALALLSLSGNTGADSLGREAKCYNELNGCTKIYDPVCGTDGNTYPNECVL
CFENRKRQTSILIQKSGPC
>sp|P00996|ISK1_BOVIN Pancreatic secretory trypsin inhibitor
MKVASIFLLTALVLMSLSGNSGANILGREAKCTNEVNGCPRIYNPVCGTDGVTYSNECLL
CMENKERQTPVLIQKSGPC
>tr|F6VUK4|F6VUK4_ORNAN OS=Ornithorhynchus anatinus GN=SPINK1 PE=3 SV=1
MRITGVFLLLLATFLCFSDLAGALGEGREPNCKTPLAKCTKIYEPVCGSDGETYANECLL
CEANKPRKEHVLVRKSGKC
>sp|P61013|PPLA_PIG Cardiac phospholamban OS=Sus scrofa GN=PLN PE=1 SV=1
MDKVQYLTRSAIRRASTIEMPQQARQNLQNLFINFCLILICLLLICIIVMLL
```

1.2.2 Sequence Format Conversion Tools

Although FASTA is a very commonly used file format, sometimes you may need to use other formats. You can use ***ReadSeq*** in order to convert text files with protein sequence among different formats.

ReadSeq— Conversion between various sequence file formats	http://www.ebi.ac.uk/Tools/sfc/readseq/	
	http://csbg.cnb.csic.es/PB/E1010	

Within ReadSeq's main form, you can paste your sequence or upload a file with it. Then select the input format of your sequence (you can also leave the default option "Auto-detected") and the output format you want to convert to. Finally click the "Submit" button.

1.3 Main Protein Sequence Databases

A protein sequence database is a collection of amino acid sequences and, in some cases, associated information. Most databases organize both, the sequence and its associated information, into an entry. What exactly constitutes an entry depends on the database. This is due to the different possible sources of variation of a protein sequence, like alternative splicing products and isoforms, natural variants, etc., as well as sequence redundancy, that is, the same sequence from different sources, either biological source (different organisms) or experimental source (different results coming from different studies). A database entry is labeled with a database identifier. A database identifier should ideally be unique and permanent, so as to unambiguously refer to the information stored for that entry (protein) in the database and serve as a mean to cite and connect it from external resources.

In the case of protein sequences, in general we can say that a database entry is a combination of a sequence (or set of related sequences) and additional data relevant to the sequence(s), that can be uniquely identified with a public identifier. We recommend reading the database documentation and/or publications to fully familiarize with its contents.

The main database containing protein sequences together with their associated functional information is the ***UniProt*** (UniProt 2014). The UniProt is in fact a group of databases. We will discuss two of them: the UniProt Knowledge Base and UniRef.

UniProt—Protein sequences and associated functional information	http://www.uniprot.org	
	http://csbg.cnb.csic.es/PB/E1020	

The **UniProt Knowledge Base** (UniProtKB), its core database, is made of two large sequence collections: the Swiss-Prot (manually annotated and reviewed sequences) and TrEMBL (automatically created –e.g. translated from genomic data- and annotated sequences). From the practical point of view, you won't probably notice if a particular entry belongs to one or the other. But for the quality of the data, sometimes is important to know if the information stored in a given protein entry has been checked by an expert.

Most sequences in the UniProtKB do not come from direct protein sequencing, but from translation of nucleotide sequences obtained experimentally. Around 98 % of the sequences are translations of coding sequences stored in DNA/RNA sequence databases. This means that you should expect a lot of sequence redundancy, variable quality (e.g. fragments), and incomplete or tentative functional annotations especially in the TrEMBL collection.

The **UniRef** contains three collections of sequences (UniRef100, UniRef90 and UniRef50), where redundant sequences have been eliminated (in fact they are not eliminated, but clustered and represented by a single sequence). These sets are particularly useful when we are not interested in considering all known sequences (including those virtually identical), but just a set of representatives of the known sequence universe. UniRef100 does not contain two identical sequences (only sequences that are less than 100 % identical), UniRef90 contains sequences that are less than 90 % identical, and UniRef50 less than 50 % identical.

1.3.1 Sequences and Database Entries

Entries in UniProtKB are uniquely identified by their accession number (AC). For example, P00995 is the accession number in UniProtKB for the protein sequence encoded in humans by the SPINK1 gene. Occasionally you might find entries with a primary accession number and additional secondary accession numbers. Secondary accession numbers are kept in order to ensure persistence (backward compatibility). You always can cite (refer to) the product of the human SPINK1 gene in UniProt providing its primary accession number: P00995.

To access the content of entry P00995 you can enter this accession number on the Search form of the UniProt web site. To download the protein sequence corresponding to P00995, just click on the "FASTA" button (close to the sequence shown in the browser) or click on the "Format" tab in the main menu and select "FASTA (canonical)".

```
>sp|P00995|ISK1_HUMAN ...
MKVTGIFLLSALALLSLSGNTGADSLGREAKCYNELNGCTKIYDPVCGTDGNTYPNECVL
CFENRKRQTSILIQKSGPC
```

Note that occasionally you might find references to UniProtKB entries not using accession numbers (AC), but entry names (ID). For example entry P00995 might be referred as ISK1_HUMAN. Some people like using entry names, especially for Swiss-Prot (as they are a combination of two mnemonics giving hints about gene name and organism). Nevertheless note that, unlike accession numbers, entry names are not stable identifiers, and as such it is not recommended to use them to refer to UniProt entries (See http://www.uniprot.org/manual/entry_name).

Each database entry might contain more than one sequence. Currently, there is a canonical protein sequence (shown as default in the UniProt web page), and an optional set of sequence isoforms (e.g. products of alternative splicing).

For example, entry Q04679 contains the sequences of two isoforms, produced by alternative splicing, of the subunit gamma of a sodium/potassium-transporting ATPase, both encoded by the rat Fxyd2 gene. The two sequences are labeled as Isoform 1 (the canonical sequence) and Isoform 2, and can be cited as Q04679-1 and Q04679-2, respectively. They differ in that, while the canonical isoform has the sequence MTELSANH in positions 1-8, isoform 2 has MDRWYL.

To download all the protein sequences of Q04679, search UniProt with that entry accession. Once in the Q04679 page, click on the "FASTA" button (close to each of the isoform sequences shown in the browser) or click on the "Format" in the main menu and select "FASTA (canonical & isoform)".

```
>sp|Q04679|ATNG_RAT
MTELSANHGGSAKGTENPFEYDYETVRKGGLIFAGLAFVVGLLILLSKRFRCGGSKKHRQ
VNEDEL

>sp|Q04679-2|ATNG_RAT Isoform 2
MDRWYLGGSAKGTENPFEYDYETVRKGGLIFAGLAFVVGLLILLSKRFRCGGSKKHRQVN
EDEL
```

Other sources of variation within a protein sequence in UniProt are natural variants and sequence conflicts. Both are annotated as sequence features (see next section), and therefore cannot be directly downloaded as a FASTA file. If you want to work with a particular variant you should download the canonical sequence and manually edit the corresponding residue.

For example, P00995 describes 3 natural variants involved in disease (Pancreatitis, hereditary). The first is a residue change from leucine (L) to phenylalanine (F) in position 12. To work with this natural variant sequence, you have to download a file with the FASTA sequence (canonical):

```
>sp|P00995|ISK1_HUMAN ...
MKVTGIFLLSALALLSLSGNTGADSLGREAKCYNELNGCTKIYDPVCGTDGNTYPNECVL
CFENRKRQTSILIQKSGPC
```

And manually edit it (with a text editor), to reflect the L12F substitution and optionally change the header line. The edited FASTA for the natural variant will be:

```
>sp|P00995|ISK1_HUMAN Natural variant (L12F)
MKVTGIFLLSAFALLSLSGNTGADSLGREAKCYNELNGCTKIYDPVCGTDGNTYPNECVL
CFENRKRQTSILIQKSGPC
```

The Annotations

As we have mentioned, in addition to the sequence(s), the UniProtKB contains additional information on proteins. Three general types of annotations will be considered:

1. Entry information (related to database organization), including the accession number, historical accessions, etc.
2. Those related to the protein and the whole sequence:
   - Names and (mostly biological) origin
   - Protein attributes: status (like complete or fragment), and evidence of protein existence

- General annotation (Comments)
- Ontologies (including keywords and Gene Ontology (Ashburner et al. 2000) annotations)
- Bibliographic References & Cross-references (to other databases)

3. Those related to particular parts of the sequence, known as **sequence annotation features**. They refer to particular residue or a region within the sequence. The numeric positions refer always to the 'canonical' sequence. These include structural or functional information, as well as natural variations and sequence conflicts that were discussed previously.

For example, entry P00995 includes a number of annotations referring to particular regions of the sequence. We have commented some natural variants associated to disease previously. Others are related to post-transcriptional modifications and molecular processing, like disulfide bonds between Cys32-Cys61, Cys39-Cys58 and Cys47-79, or the signal peptide from residue 1 to 23.

Searching the UniProt

There are three typical scenarios to search for data in the UniProt:

1. You want to find the sequence and/or functional data for a single protein
2. You want to get the data for a list of primary accessions
3. You have a protein sequence and want to find it, as well as its sequence relatives, in UniProt

Case 1: If you want to search for a particular protein, you can do a basic or advanced search, using the search form at the top of the UniProt pages (look for the magnifying glass icon/search button). First select the UniProt database you want to search (UniProtKB, UniRef, …) and provide one or several search terms. In the basic search (the default search in the home page) terms are introduced without specifying to which field they refer to, and the search is performed across all the annotations. To do an advanced search click on "Advanced", and then construct a complex search by combining pairs term-field/section (e.g. organism, gene name, protein name, etc.). Click on "All" to expand a menu to select subsections, introduce the term you want to search, and combine with new terms and sections with the **AND**, **OR**, **NOT** operators.

Case 2: If you want to retrieve the data corresponding to a list of primary accessions, click the "Upload lists" tab on the main menu, paste the list of accessions or upload a file with them, and finally press the "Go" button. You can now browse through the results in the table, or download the data for further processing by clicking "Download". E.g. you can download all FASTA files by selecting "FASTA (canonical)" to get only canonical sequences or "FASTA (canonical & isoforms)" to get both canonical and all isoforms.

Case 3: You can search for entries with sequences identical or similar to a given sequence by clicking the "BLAST" tab on the main menu. To learn more about BLAST please refer to Sect. 1.6.

1.4 Basic Sequence-Based Characteristics

Apart from its main utility in functional and evolutionary studies (next sections), the raw sequence of a protein can be used to extract some useful properties. In this section you will get familiar with some tools for obtaining some basic features of a protein from its primary sequence.

The ***ProtParam*** server at the ExPASy Bioinformatics Resource Portal (Gasteiger et al. 2005) calculates several global parameters from the primary sequence: molecular weight, theoretical pI, amino acid composition, atomic composition, extinction coefficient, estimated half-life, instability index, aliphatic index and grand average of hydropathicity.

ProtParam—Calculation of various physico-chemical parameters of proteins	http://web.expasy.org/protparam/	
	http://csbg.cnb.csic.es/E1030	

You have to provide the UniProt accession number or paste the plain amino acid sequence, and then click on "Compute parameters". Note: FASTA format is not allowed, so remove the header line (starting with ">").

For example, for the sequence corresponding to UniProt entry P26678

```
MEKVQYLTRSAIRRASTIEMPQQARQKLQNLFINFCLILICLLLICIIVMLL
```

ProtParam produces the following output:

```
Number of amino acids: 52
Molecular weight: 6108.5
Theoretical pI: 9.50

Amino acid composition:
Ala (A)   3       5.8%
Arg (R)   4       7.7%
Asn (N)   2       3.8%
Asp (D)   0       0.0%
.....

 (B)   0    0.0%
 (Z)   0    0.0%
 (X)   0    0.0%
Total number of negatively charged residues (Asp + Glu): 2
Total number of positively charged residues (Arg + Lys): 6

Atomic composition:

Carbon      C          276
Hydrogen    H          471
...

Formula: C276H471N73O69S6
Total number of atoms: 895

Extinction coefficients:

This protein does not contain any Trp residues. Experience shows that
this could result in more than 10% error in the computed extinction
coefficient.
Extinction coefficients are in units of  M-1 cm-1, at 280 nm measured in
water.

Ext. coefficient     1615
Abs 0.1% (=1 g/l)   0.264, assuming all pairs of Cys residues form cystines

Ext. coefficient     1490
Abs 0.1% (=1 g/l)   0.244, assuming all Cys residues are reduced

Estimated half-life:

The N-terminal of the sequence considered is M (Met).
The estimated half-life is: 30 hours (mammalian reticulocytes, in vitro).
                            >20 hours (yeast, in vivo).
                            >10 hours (Escherichia coli, in vivo).

Instability index:

The instability index (II) is computed to be 65.43
This classifies the protein as unstable.

Aliphatic index: 151.92
Grand average of hydropathicity (GRAVY): 0.835
```

The ***ProtScale*** tool, also at the ExPASy Bioinformatics Resource Portal (Gasteiger et al. 2005), represents a profile along the protein sequence produced by a given numeric scale assigned to each type of amino acid. This tool can be used to have a first glimpse of the protein regions enriched in certain properties (e.g. hydrophobic domains) (Fig. 1.1).

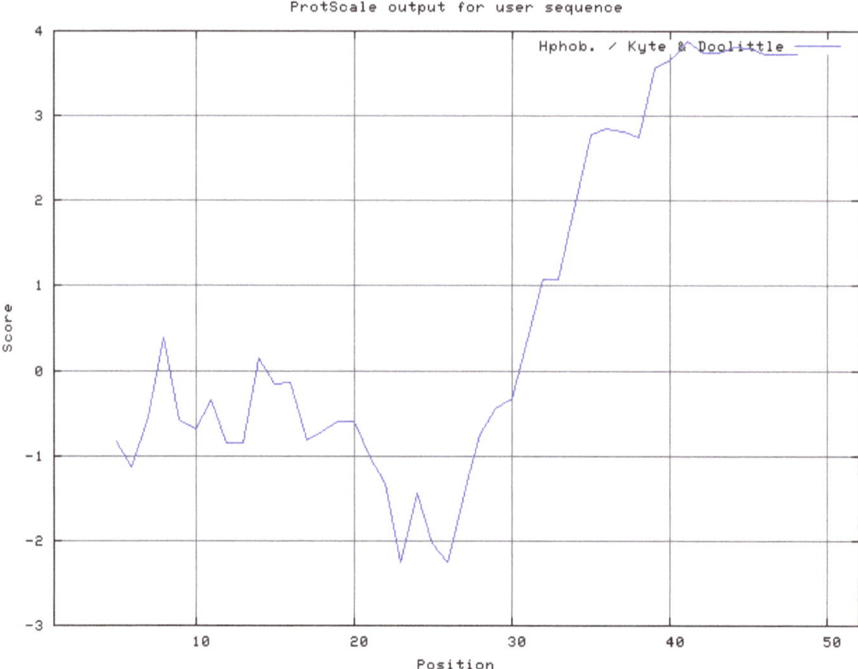

Fig. 1.1 Hydrophobicity profile for P26678 generated with ProtScale. This profile clearly shows the presence of a hydrophobic domain from residue 30 to the end. Indeed, this region is a transmembrane helix (Sect. 2.4.2)

ProtScale—Graphical representation of a property profile along the protein sequence	http://web.expasy.org/protscale/	
	http://csbg.cnb.csic.es/E1040	

For example, to visualize the hydrophobicity pattern of a sequence, enter the Uniprot accession number or paste the protein sequence, then select a hydrophobicity scale, for example "Hphob./Kyte & Doolittle", and click the "Submit" button.

1.5 Compare Two Protein Sequences

The most common strategy to compare two proteins involves aligning their sequences. An alignment of two protein sequences (also known as pair-wise or binary alignment) is a linear one-to-one correspondence of their residues. These correspondences are supposed to represent evolutionary equivalent residues: those arisen from the same ancestral residue in the ancestor of the two proteins. Evolutionary insertions/deletions of residues result in residues of one protein without equivalent ones in the other ("gaps" in the alignment).

An alignment is graphically represented by stacking together (usually in the same column) these equivalent residues and using additional characters for the gaps ('-' or '.'). In this way the correspondence between the residues of both proteins is visually evident.

Example of two sequences aligned without gaps:

```
MKVTGIFLLSALALLSLSGNTGADSLGREAKCYNELNGCTKIYDPVCGTDGNTYPNECVLCFENRKRQTSILIQKSGPC
MRITGVFLLLLATFLCFSDLAGALGEGREPNCKTPLAKCTKIYEPVCGSDGETYANECLLCEANKPRKEHVLVRKSGKC
```

Example of two sequences aligned with gaps:

```
MKVTGIFLLSALALLSLSGNT---GADSLGREAKC-YNELNGCTKIYDPVCGTDGNTYPNECVLCFENRKRQTSILIQKSGPC
MKAVGIFLLLFLAICFYQGDAEPDGAADQGTEANCGNYDLRKGCTKIFDPICGTDDVLYSNECLLCSQNLQRHTNVRIKHRGKC
```

How do programs construct an alignment? They basically try to find the alignment that maximizes a parameter known as score. This score is calculated by a scheme mostly consisting of two parts:

- Pairing of two amino acids: an amino acid scoring matrix is used. Scoring matrices are constructed empirically from the frequencies of residue substitutions observed in manually curated alignments. Example: BLOSUM62 (Henikoff and Henikoff 1992). Pairing identical or "similar" amino acids (those frequently substituting each other during the evolutionary process, e.g. E-D) contributes positively to the score while "dissimilar" amino acids (e.g. E-V) contributes negatively.
- Pairing of an amino acid and a gap: a gap penalty scheme, which contributes negatively to the score by penalizing the number and length of the gaps.

The final score is obtained by accumulating the partial scores for each pair of equivalent positions. Hence this score increases if the same or frequently substituted amino acids are paired, and decreases when pairing amino acids with gaps or with other amino acids whose substitution is rarely observed.

For most of the cases, alignment programs work fine with the default options for these parameters, so you don't need to change them. If you want to change these default values, search for the "Advance/More options" control.

A sequence alignment allows not only to visually compare two sequences, but to quantify their similarity as well. So, in addition to the two sequences aligned, programs that build pairwise alignments provide also quantitative data, most commonly:

- Identity: percentage of exact residue matches (number of matches divided by alignment length).
- Similarity: percentage of conservative matches, computed based on a scoring matrix.

It is important to remember that identity and similarity are calculated for a particular alignment. Therefore, a different alignment of the same two sequences will, in general, provide different values of identity and similarity.

Most programs do not assume a particular relationship between the two sequences when building the alignment. Therefore, the exact meaning of the (global and local) correspondences implicit in an alignment depends on our expert judgment. In general, a high similarity might imply an evolutionary, structural or functional relationship between the two proteins. But remember: pairwise sequence alignment programs always generate an alignment for any two sequences, regardless of whether they are actually related or not.

1.5.1 Types of Pair-Wise Sequence Alignments

You can perform two types of sequence alignments, depending on the type of relationship you expect between the two sequences: **Global alignment:** tries to align the entire sequences of both proteins (Needleman and Wunsch 1970). **Local alignment:** tries to align only the most similar parts (segments) of the two proteins (Smith and Waterman 1981).

Both methods are accessible through the ***EMBOSS*** suite at the EBI (Rice et al. 2000).

EMBOSS Needle—Global alignment of two protein sequences	http://www.ebi.ac.uk/Tools/psa/emboss_needle	
	http://csbg.cnb.csic.es/PB/E1050	

EMBOSS Water—Local alignment of two protein sequences	http://www.ebi.ac.uk/Tools/psa/emboss_water	
	http//csbg.cnb.csic.es/PB/E1060	

For both programs, paste the two protein sequences or upload files with them (multiple formats supported, including FASTA) and click the “Submit” button. You can display/change the default options by clicking on “More options…”.

In case the two sequences are very similar, the results provided by the two programs will be also similar. Otherwise, results will differ. As a general rule, use global alignment (EMBOSS Needle) when you want to evaluate whether two sequences are overall evolutionary related; and local alignment (EMBOSS Water) when you expect the similarity to be restricted to a particular region (motif, shared domain, etc.).

As an example, we are going to compare the results produced by both programs for the following two sequences:

```
>tr|Q73YB7|Q73YB7_MYCPA (frag)
EGARMLYIHPDECVDCGACEPVCPVESIYYEDDLPPEHSQYLQIN

>sp|O58412|VORD_PYRHO (frag)
KYCPEPAIYIKEDGFVAIDYDYCKGCGICANECPTKAITMVREEK
```

Result from EMBOSS-Needle:

```
# Length: 59
# Identity:      11/59 (18.6%)
# Similarity:    16/59 (27.1%)
# Gaps:          28/59 (47.5%)
# Score: 39.0

Q73YB7_MYCPA       1 ------------EGARMLYIHPDECVDCGACEPVCPVESI--YYEDDLPP      36
                                 :|  .:.|..|.|..||.|...||.::|  ..|:.
VORD_PYRHO         1 KYCPEPAIYIKEDGFVAIDYDYCKGCGICANECPTKAITMVREEK---      45

Q73YB7_MYCPA      37 EHSQYLQIN     45

VORD_PYRHO        46 ---------     45
```

Result from EMBOSS-Water:

```
# Length: 18
# Identity:       8/18 (44.4%)
# Similarity:    10/18 (55.6%)
# Gaps:           0/18 ( 0.0%)
# Score: 50.0

Q73YB7_MYCPA      11 DECVDCGACEPVCPVESI      28
                     |.|..||.|...||.::|
VORD_PYRHO        21 DYCKGCGICANECPTKAI      38
```

In the alignments generated by these two (as well as other) programs, special characters are used to highlight the equivalences between the positions of the two sequences:

- '|' denotes an exact match
- ':' denotes a conservative match
- '.' denotes a non-conservative match

Note that for exactly the same two sequences, two different alignments have been constructed. With the first (global approach) the identity is 18.6 %, and with the second (local approach) identity is 44.4 %. This is due to the fact that identity is calculated respect to the total length of the aligned region (being 59 in the global, and 18 in the local), and that the number of exact matches that also differ (11 in the global, and 8 in the local).

1.6 Finding Similar Sequences in a Database (Basic)

You can find information in protein sequence databases by text-based searches, just like in any other web-accessible database. Nevertheless, in this section we show you a most powerful way to search information on proteins, that is, using their unique biochemical fingerprint: their sequences.

Finding the set of sequences in a database similar to one of interest is the first step in most protein bioinformatics studies. This is because similar sequences are expected to represent homologous proteins, that is, proteins with the same evolutionary origin. As you will see in the next sections, comparing a group of homologous proteins provides important information on them. Homologous proteins have the same global three-dimensional structure and share many functional characteristics. But not only the similarities among homologous proteins are rich sources of information, but the differences as well.

From a practical point of view, there are two main ways to search a database for similar sequences:

1. By pair-wise comparisons, i.e. the one you provide (query sequence) and every other sequence in the database. This basic way of searching is the subject of this section.
2. By comparing a sequence and a set of sequences (mostly in the form of a multiple sequence alignment). This type of search is the subject of the Sect. 1.8.

To search for similar sequences, you will start with a **query sequence**, and then you need to decide which database and program you are going to use to actually do the search. Finally, consider whether you should worry about advanced options or not.

1.6.1 Which Sequence Database to Search?

That depends very much on what are you want to do with the protein sequences found. If you are interested in:

- Obtaining a representative set of the sequence universe around your query sequence, then consider searching against a non-redundant database (like UniRef databases)
- Inferring functional information for your protein (transferred from that associated to the homologs), then consider searching against a sequence database with functional info (like UniProt, especially the Swiss-Prot subsection)
- Retrieving structural data, then consider searching against a sequence database with proteins of known structure (like the PDB, see Sect. 2.2).
- Finding homologs in particular organisms, then consider searching against genome/proteome-specific databases, or filter by taxonomy (if this functionality is available)
- ....

As we have mentioned, basic sequence-based searches are in fact a set of successive pair-wise alignments of the query sequence with each sequence in the database. Therefore, independently of the program and database you use, most sequence searches will return:

- A list of hits, similar sequences (potential homologs) found in the database, ordered by relevance
- A pairwise alignment for each hit, comparing the query sequence and the hit sequence

1.6.2 BLAST

The most widely used program to find similar sequences in large databases is ***BLAST* (Basic Local Alignment Search Tool)** (Altschul et al. 1990). BLAST uses a heuristic approach for performing the pairwise alignments, based on the matching of short sequence fragments, instead of the alignment methods commented in the previous section. This approach accelerates the process by several orders of magnitude while rendering very similar results to the exhaustive methods and, consequently, it is the preferred choice for scanning large sequence databases.

BLAST is in fact a family of programs. You can use three of them to do basic protein sequence-related searches:

- To search for similar protein sequences to a query protein sequence (**blastp**)
- To search for protein sequences related to a query DNA/RNA sequence (**blastx**)
- To search for nucleotide sequences related to a query protein sequence (**tblastn**)

There are many blast servers available online. In the following we are going to comment that at the NCBI of the NIH.

NCBI BLAST—Search for similar protein sequences in databases	http://www.ncbi.nlm.nih.gov/BLAST	
	http://csbg.cnb.csic.es/PB/E1070	

To search for sequences similar to your protein of interest (query protein), follow the "protein blast" link within the "Basic BLAST" section. Then paste your sequence (or alternatively upload a file containing it) and select the Database to search. Optionally you can restrict the search to an organism or a taxonomy group. **Blastp** (protein-protein BLAST) is the default "Program Selection" (the type of search we want to do right now). Finally click the "BLAST" button to proceed.

Other protein-related database searches that can be done with BLAST are:

- If you want to search for DNA/RNA sequences that code for proteins similar to yours, use **tblastn**. That will search your query protein sequence against translated nucleotide sequences. In this case BLAST will automatically translate nucleotide sequences in the database using the six possible reading frames. This might be useful, for example, to find genomic regions potentially coding for homologs of your protein but that have not been annotated as genes yet.
- If you have a nucleotide sequence (DNA or RNA), and want to search in a protein database for similar sequences to those coded in any of the six frames of the query, then follow the **blastx** link in the "Basic BLAST" section. This option is useful, for example, to retrieve the homologs of the product(s) of a gene for which you are not sure of the intron/exon structure (Indeed, the protein homologs eventually found will map on the exons of your gene).

The defaults for the other BLAST input parameters work fine for most searches and usually you will not have to touch them. A parameter which you might need to play with in certain circumstances is the "Filter" (click "Algorithm parameters" at the bottom of NCBI BLAST search form). By default, BLAST filters are intended to ignore some sequence segments with particular amino acid compositions which might cause problems in the sequence comparisons behind the search. These are usually "non informative" segments with, for example, repeated amino acids, and are substituted by "X" ("masked") in the forthcoming operations and in the display of the results. Nevertheless, some particular types of proteins have this kind of motifs and with a functional reason (e.g. un-structured proteins, Sect. 2.4.4). If we suspect that these masked regions might be important and conserved in the homologs we are looking for, we may try a search with this filter turned off. But, again, leaving it on is ok for most situations.

In addition to the NCBI BLAST site, most protein sequence databases available on the Internet allow you to perform a BLAST search against their contents. This is the case of the BLAST search in UniProt website (see Searching the UniProt).

Interpreting BLAST's Output

As an example to explain a typical BLAST output, we are going to look for homologs of known three-dimensional structure for the *Rhodococcus opacus* 7Fe ferredoxin (Uniprot: C1BDF9). That is, searching for similar sequences in the PDB (Sect. 2.2).

```
>C1BDF9 7Fe ferredoxin
MAFVIGEPCVDVMDKSCIEECPVDCIY
EGGRMLYIHQNECIDCGACEPVCPVSAIRPAQKVDEQWQPFQESSREVTKDL
GSPGGSSNVDLPIADSEYVKTLDEK
```

For that, in the NCBI BLAST web form we paste this sequence and select "Protein data bank (PDB)" in the "Database" section.

At the top of the results page, we find a graphical representation of the found hits indicating their coverage respect to the query sequence and the score of the alignments with a color scale (Fig. 1.2).

Below this graphical overview, you can find the list of hits (found sequences) ordered by the significance of the match, with the most significant at the top. The significance is measured by a parameter called Expected Value (E-value). The E-value measures the probability of obtaining such a good hit when searching a database of a particular size just by chance. So the closer to zero the E-value, the more significant the hit will be.

As a rule of thumb, hits with E-values lower than 10^{-4}–10^{-5} can be considered clear homologs, while relaxing to 10^{-3} we will still find true homologs, but the proportion of false positives will increase as well. The percentage of identity between the query and the hit is also indicated. Apart from the E-value and the other parameters, we recommend you to visually inspect the alignments between the query sequence and each of the hits you are going to consider. These alignments are shown just below the list of hits in this particular BLAST server (NCBI). Clicking the description of the protein in the hits list jumps directly to the corresponding alignment

As commended previously, in order to perform searches against large databases in a reasonable time BLAST uses a heuristic approach to speed up the process of aligning two sequences. Using a heuristic means in practical terms that BLAST is not exhaustive in trying to find the optimal alignment but works well enough to find relevant similar sequences. Therefore if you want to find similar sequences in a database, you can use it. But if you want to obtain the optimal alignments between your query and the found hits, don't rely on the alignment provided by BLAST unless the two proteins are very similar. Instead, take the sequences you have found with BLAST and perform a pair-wise (Sect. 1.5) or multiple sequence alignment (Sect. 1.7) with a non-heuristic method to obtain the optimal alignment.

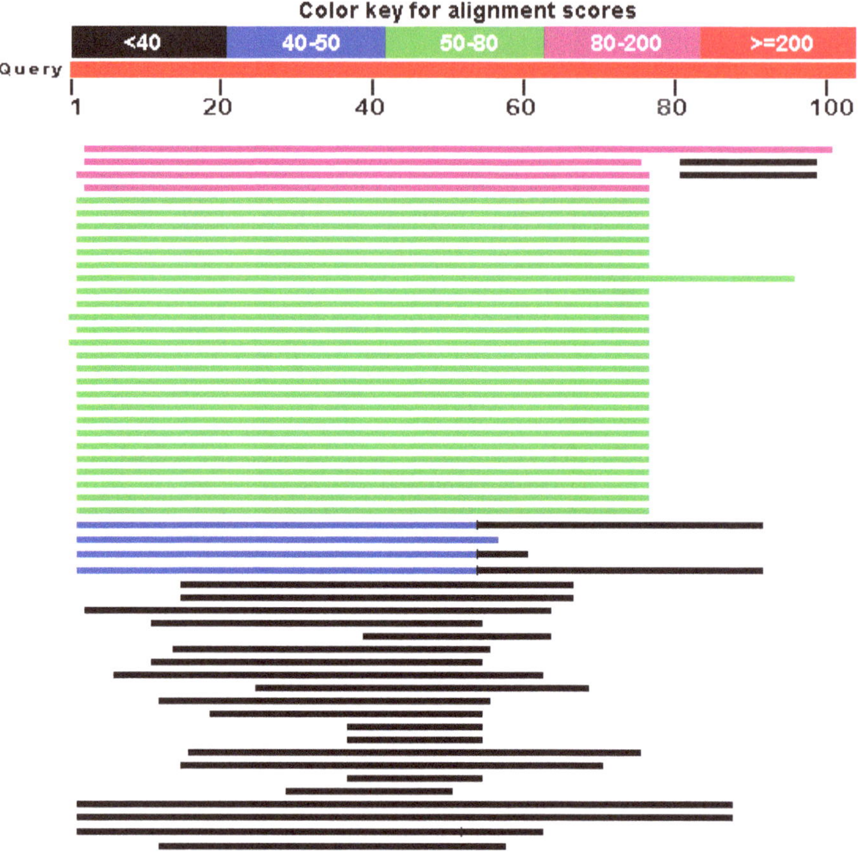

Fig. 1.2 Graphical representation of the C1BDF9 homologs found in PDB by BLAST

To save the sequences of the hits you are interested in, select them in "Sequences producing significant alignments" list with the checkboxes on the left. Then click the Disk icon (Download). A pull-down menu appears. You can download the complete sequences of the hits (in FASTA or GenBank formats), or alternatively only the region of the hit sequences that are aligned to the query sequence (in FASTA format).

As BLAST uses a local alignment approach do not expect hits to cover the full length of your query sequence. And remember that a hit may be in fact a segment of a longer sequence that aligns only partially to your query sequence.

In the example above, the best hit PDB 2V2K (chain A) (length:105) is significant (E-value 2e-35, sequence identity: 58 %) and covers 95 % of the length of the query sequence. It corresponds to the crystal structure of Fdxa, a 7fe ferredoxin from another bacteria (*Mycobacterum smegmatis)*.

C1BDF9-2V2K_A BLAST alignment:

```
Query  3    FVIGEPCVDVMDKSCIEECPVDCIYEGGRMLYIHQNECIDCGACEPVCPVSAIRPAQKVD  62
            +VI EPCVDV DK+CIEECPVDCIYEG RMLYIH +EC+D GACEPVCPV AI     V
Sbjct  2    YVIAEPCVDVKDKACIEECPVDCIYEGARMLYIHPDECVDXGACEPVCPVEAIYYEDDVP  61

Query  63   EQWQPFQESSREVTKDLGSPGGSSNVDLPIADSEYVKTL  101
            +QW  + +++ +   +LGSPGG+S V      D + +K L
Sbjct  62   DQWSSYAQANADFFAELGSPGGASKVGQTDNDPQAIKDL  100
```

Note that BLAST highlights the identities in its alignments with the corresponding aminoacid, and the similarities with a "+".

The second hit, 1H98 (chain A) (length: 78), is also significant (E-value 8e-25, seq. id: 57 %), but only covers 71 % of the length of the query sequence.

```
Query  3   FVIGEPCVDVMDKSCIEECPVDCIYEGGRMLYIHQNECIDCGACEPVCPVSAIRPAQKVD  62
            VI EPC+ V D+SC+E CPV+CIY+GG   YIH  ECIDCGAC P CPV+AI P + V
Sbjct  2   HVICEPCIGVKDQSCVEVCPVECIYDGGDQFYIHPEECIDCGACVPACPVNAIYPEEDVP  61

Query  63  EQWQPFQESSREVT  76
           EQW+ + E +R++
Sbjct  62  EQWKSYIEKNRKLA  75
```

Hit 1G3O (chain A) is 106 residues long, but only a partial segment (residues 1–76) is similar to the query sequence.

```
Query  2   AFVIGEPCVDVMDKSCIEECPVDCIYEGGRMLYIHQNECIDCGACEPVCPVSAIRPAQKV  61
           AFV+ + C+      C+EECPVDC YEG   L IH +ECIDC  CEP CP  AI    +V
Sbjct  1   AFVVTDNCIKCKYTDCVEECPVDCFYEGPNFLVIHPDECIDCALCEPECPAQAIFSEDEV  60

Query  62  DEQWQPFQESSREVTK  77
            E  Q F + + E+ +
Sbjct  61  PEDMQEFIQLNAELAE  76
```

If that search was performed, for example, to look for a suitable template to construct a 3D model of C1BDF9 (Sect. 2.5.1), we would take the first hit (2V2K_A) since it covers most of the sequence.

NCBI BLAST generates other interesting reports on the search results that can be accessed through the "Other reports" links at the top. The "Taxonomy reports" link provides three tables based on the information available for the hits' organisms in the NCBI Taxonomy Database: "Lineage" (focuses on the organism of the first hit), "Organism" (shows the hits grouped by organism) and "Taxonomy" (summarizes the main taxonomic groups and number of hits found).

You can also align all the hit sequences through the "Multiple alignment" tag in "Other reports" menu, or a subset of them by selecting with the checkboxes in the "Descriptions" section and follow the "Multiple alignment" link that shows up at the top of this section. NCBI BLAST uses COBALT (Constraint-based Multiple Alignment Tool) (Papadopoulos and Agarwala 2007) to generate the multiple sequence alignment (see Sect. 1.7 for alternative programs to use).

1.7 Compare More than Two Sequences

If you want to compare more than two sequences then you need a program to build a **multiple sequence alignment** (MSA). In a multiple sequence alignment, proteins are arranged in such a way that equivalent residues are piled up together (e.g. in the same column). In such representation, a column (set of evolutionary equivalent residues) is also called "position".

Multiple sequence alignments of homologous proteins are rich sources of functional and structural information since they can be seen as representations of the mutations allowed by the evolution at each position. They can be used for finding positions related to functionality (Pietrosemoli et al. 2012), such as conserved positions, as well as for generating profiles than can serve for fishing distant members of the family (Sect. 1.8).

Example of a multiple sequence alignment of 5 proteins:

```
PSRARRDAVG--DH--PAVEALP----PQSGPHKKEISFFTVRKEEAADADLWFPS
PGGASK--VGQTDNDPQAIKDLP----PQGED------------------------
PGGAAK--VGKVANDPQFVKALP----QQAAHE-----------------------
PGGSSN--VDLPIADSEYVKTLD----EK---------------------------
PGGAAK--LGRVGADTPLVAALPASTTPHP--------------------------
```

Programs that build MSAs use more complex methods than those that compare only two sequences, and therefore require more time to run. Like in the case of sequence database searches, MSA methods use heuristics rather than global optimization.

In practical terms this means that there are various programs using different strategies that will provide, in general, slightly dissimilar results. It is therefore advisable, in the most difficult cases (proteins not very similar) to try several programs, to visually inspect the results, and manually correct the alignment in case it is needed. You can also try to use additional information to correctly build the alignment, such as structural data.

Be aware of possible restrictions in the length or number of sequences to align when using the Web-based programs commented below.

Clustal Omega (Sievers et al. 2011) is the latest addition to the Clustal series of programs for building multiple sequence alignments. It offers a good trade-off between speed and accuracy for building MSAs with a large number of sequences.

Currently, the Clustal Omega web server at the EBI is limited to 2000 sequences and 2 MB of data. If you want to align more sequences (or lager datasets) then you should install Clustal Omega locally (see http://www.clustal.org/omega/).

Clustal Omega—Multiple alignment of protein sequences	http://www.ebi.ac.uk/Tools/msa/clustalo	
	http://csbg.cnb.csic.es/PB/E1080	

In the main form, paste or upload the set of sequences to align (e.g. a FASTA file containing all unaligned sequences) and select the output format (or leave the default 'Clustal w/o numbers'). Optionally you can provide an e-mail address to be notified in case of long runs. The rest of parameters ("Advanced options") are fine for most situations. Finally click the "Submit" button.

The following is an example alignment produced by Clustal Omega, in Clustal format

```
CLUSTAL O(1.2.1) multiple sequence alignment


UniRef50_A4T0N6      --------MQSRRDQTRDFRPLTTTAARPTVDGMTYVIGSACVDIVDKSCVQECPADCIY
UniRef50_W0RJQ2      ---------------------------------MPYVITEACINVKDKSCVDVCPVDCIY
UniRef50_C1BDF9      ---------------------------------MAFVIGEPCVDVMDKSCIEECPVDCIY
UniRef50_Q9FC76      ---------------------------------MTYVIAQPCVDIKDRACVTECPVDCIY
UniRef50_P00215      ----------------------------------TYVIAEPCVDVKDKACIEECPVDCIY
UniRef50_W5W0F2      MAGAGQGILTVRREP--LMNVLGRMRKREQETAVTYVIAEPCVDVLDKACIEECPVDCIY
                                                        :** . *::: *::*:  **.****

UniRef50_A4T0N6      EGDRAMYINPNECVDCGACRIACRVDAIYYETDLPDEELAFLDDNAAFFTTTLSGRDEPL
UniRef50_W0RJQ2      EGPEQLFIHPDECIDCGACEPECPVTAIFPDEDVPAQLRDYVRKNREVFQ--------SD
UniRef50_C1BDF9      EGGRMLYIHQNECIDCGACEPVCPVSAIRPAQKVDEQWQPFQESSREVTK--------DL
UniRef50_Q9FC76      EGARTLYINPAECVDCHACEPVCPVEAIFHEDDLPRHWAHYLAVNAEYFD--------EA
UniRef50_P00215      EGARMLYIHPDECVDCGACEPVCPVEAIYYEDDVPDQWSSYAQANADFFA--------EL
UniRef50_W5W0F2      EGDRMLYIHPDECVDCGACEPVCPVEAIYYEDDVPDQWKDYTKANADFFD--------EL
                     ** . ::*.  **:** **.  * * **    .:  .   :   .

UniRef50_A4T0N6      GDPGGAAKLGRVGADTPLVAALPASTTPHP----------------------
UniRef50_W0RJQ2      EPPGRPTR--------------------------------------------
UniRef50_C1BDF9      GSPGGSSNVDLPIADSEYVKTLDEK---------------------------
UniRef50_Q9FC76      ASPSRARRDAV--GDHPAVEALPPQSGPHKKEISFFTVRKEEAADADLWFPS
UniRef50_P00215      GSPGGASKVGQTDNDPQAIKDLPPQGED------------------------
UniRef50_W5W0F2      GSPGGAAKVGKVANDPQFVKALPQQAHE------------------------
                       *.   .
```

In CLUSTAL alignments, the last line includes special characters to highlight the conserved positions:

- '*' denotes an exact match
- ':' denotes a conservative match
- '.' denotes a less conservative match

Clustal Omega is also the program used in the UniProt web site (Sect. 1.3) to align sequences. To use it within this site go to the "Align" tab in the main menu in order to paste the sequences (in FASTA format) or a list of UniProt identifiers, and then press "Run Align". Alternatively, you can also select a set of sequences from the tabular output of any type of UniProt search (either text or BLAST search) and then press the "Align" button at the top of the table.

The resulting alignment can be conveniently colored by UniProt sequence feature annotations or amino acid properties (select those from the "Highlight" panel on the left). To download the alignment file, click the "Download" button at the top of the page.

Another widely used program for multiply aligning is ***MAFFT*** (Katoh and Standley 2013). The peculiarity of this program is that it offers different pre-defined settings where different multiple sequence alignment strategies are implemented. These represent different trade-offs between speed and accuracy, adequate for different situations and requirements.

Among those oriented toward accuracy, there are three strategies recommended when you want to align less than 200 sequences:

- E-INS-i: if your proteins are expected to have multiple conserved domains and long gap regions
- L-INS-i: if your proteins are expected to have only one conserved domain and long gap regions
- G-INS-i: if your proteins are expected to be globally similar.

MAFFT—Multiple alignment of protein sequences	http://mafft.cbrc.jp/alignment/server/	
	http://csbg.cnb.csic.es/PB/E1090	

Within MAFFT's web form, paste or upload your protein sequences (e.g. a FASTA file containing all unaligned sequences), optionally provide an e-mail address to be notified of job completion, select the "Strategy" to follow (see above), and finally click the "Submit" button.

The ***Expresso*** program (Armougom et al. 2006) of the T-coffe suite (Notredame et al. 2000) aligns multiple protein sequences using the information contained in homologous proteins of known structure. Therefore it is a good option to align protein sequences when the structure of at least two of them (or close homologs) have been solved.

Expresso—Multiple sequence alignment guided by structural information	http://tcoffee.crg.cat/apps/tcoffee/do:expresso	
	http://csbg.cnb.csic.es/PB/E1100	

In the form, paste or upload a file with the sequences to align (in FASTA format), provide an email address (optional, if you want to receive a copy of the results by e-mail) and click the "Submit" button.

The resulting MSA is conveniently colored to reflect the quality of the alignment in the different regions, in a color scale from blue (bad) to red (good).

1.7.1 Multiple Sequence Alignments: Formats and Conversion

There is a wide variety of file formats to store multiple sequence alignments. As we have mentioned in Sect. 1.2.1.1, FASTA can also store multiple sequence alignments. Most programs for generating and handling multiple sequence alignments can read and produce FASTA files. In case you need to work with other MSA formats, you can use tools such as ***myhits reformat MSA*** to convert from FASTA to other multiple sequence alignment formats.

Myhits reformat MSA—Conversion between multiple sequence alignment file formats	http://myhits.isb-sib.ch/cgi-bin/reformat	
	http://csbg.cnb.csic.es/PB/E1110	

In the main form, paste as input your MSA in FASTA format, select the output format(s) from the options provided, and click the "Reformat" button.

1.7.2 Alignment Editing and Representation

Alignment visualization and editing tools allow you to visually and interactively inspect the MSA as well as editing it (e.g. to manually correct misaligned regions). They also allow to visually obtain a first insight into functional features (conserved regions, etc.). In these tools, you can also find additional functionalities, such as adding annotations to aligned residue positions: typically secondary structure features, or any label you consider relevant (active site, mutations, etc.). These tools also allow coloring the alignment according with different schemas and provide editing capabilities for sorting, finding and deleting sequences, and removing or hiding certain blocks (consecutive portions).

Jalview (Waterhouse et al. 2009) is a widely used program for visualizing and editing sequence alignments.

Jalview—Visualization and editing of multiple sequence alignments	http://www.jalview.org	
	http://csbg.cnb.csic.es/PB/E1120	

Jalview is a program written in Java, that can be run as an Applet embedded in a web page (see general comments on running Applets in Sect. 2.3.1) or launched as a desktop application directly from a Web browser (in this case, you should have the Java Runtime Environment installed in your computer, see www.java.com for further information). In this book, we briefly describe how to work with the desktop application which automatically opens from the web browser.

To start Jalview click the "Launch Jalview Desktop" button in Jalview's Home Page and close all the example panels that show up within Jalview main window (from the main menu, select "Window → Close all").

You can now open a MSA file from the main menu ("File → Input Alignment → From File") or you can paste a MSA that is already opened in a text editor or web browser ("File → Input Alignment → From TextBox").

Assigning colors to your alignment helps you to check the quality of the alignment as well as to display some properties of the different regions. From the main menu choose "Colour → Clustalx". See http://www.jalview.org/help/html/colourSchemes/clustal.html for a full description of this color scheme. Other amino acid color schemas intended for highlighting different features of the sequences are available.

In addition to the alignment itself, Jalview allows you to visualize annotations associated to its positions. Three rows of quantitative annotations are automatically generated by Jalview once an alignment is loaded: Conservation (of physicochemical properties), Quality (using an ad hoc measure based on BLOSUM62 scores) and Consensus (most frequent residue, and corresponding frequency). Please, refer to Jalview Help pages for the description of methods used to calculate these parameters.

To manually add a new annotation line, right-click on the annotation labels on the left and select "Add new row". Introduce a name and a short description. A new annotation line appears. Then, select the position or range of positions you want to annotate along the alignment, right-click and select the type of annotation from the drop-down menu, most general being "Label". Helix and Sheet are specifically used to represent secondary structure elements as red bars (helix) and green arrows (sheets) (Fig. 1.3).

To select a sequence or a set of sequences, click the sequence ID(s) shown on the left (use Ctrl and Shift keys for multiple selections). You can delete sequences once selected, pressing the Backspace key or from the menu: "Edit → Delete". You can also change the order in which sequences are shown: select the sequence(s) you want to move and use the up and down arrows from the keyboard.

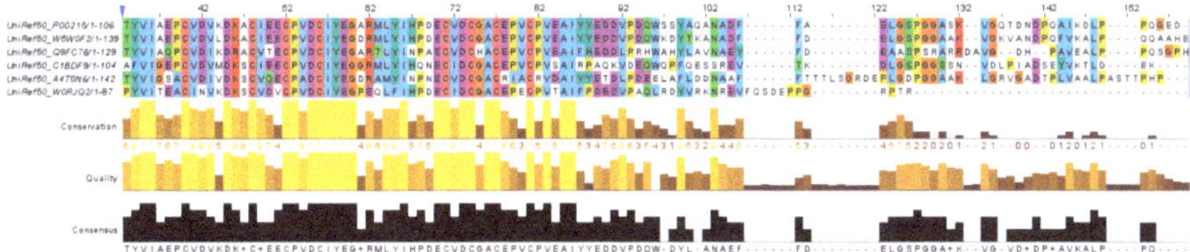

Fig. 1.3 Representation of an alignment in Jalview, with different quantitative features associated to the positions (*bars*)

To select a column or a block of columns in the alignment, click on the column numbers (use Ctrl and Shift keys for multiple selections). You can hide selected columns by right-clicking on selection. You can delete selected columns pressing the Backspace key or through the menu with "Edit → Delete".

One of the most useful functionalities of Jalview is the possibility to manually adjust the alignment, through the insertion or deletion of gaps:

- To add or delete gaps within an individual sequence, press and keep pressed the Shift key and click and drag from the position you want to adjust (right to add gaps and left to delete).
- To add or delete gaps for a group of sequences, select those sequences (as explained above), press and keep pressed the Ctrl key and click and drag from the position you want to adjust as in the case of an individual sequence

Remember that you can always undo your last action with "Edit → Undo" (or Ctrl + Z).

Other commonly used actions in Jalview related to alignment edition are:

- "Edit → Remove empty columns" to delete positions that contain only gaps.
- "Edit → Remove redundancy" to keep only sequences with a percentage of identity lower than a given threshold.
- "Format → Wrap" to show the alignment in blocks of a given length. This is useful for printing, generating figures, etc.

To save the alignment select "File → Save" or "Save as…" (to save with another name or format). Specify the format in the "File format" drop-down menu. Several MSA formats are supported (including FASTA and Clustal). If you want to save also the annotations select "Jalview.jar" format.

To save a whole project (which basically stores the working session for a future use), from the main window menu select "File→Save project". Jalview projects have a ".jvp" extension.

To get additional help on Jalview, from the main menu select "Help → Documentation" (or press F1).

Although less featured than Jalview, the ***Alignment-Annotator*** (Gille et al. 2014), hosted at the Bioinformatics Organization Inc. has the advantage that it runs entirely within a web browser. Consequently, it is a good alternative if we do not have the Java Runtime Environment (JRE) installed in our system.

Alignment-Annotator—Online visualization and manipulation of MSAs	http://www.bioinformatics.org/strap/aa/	
	http://csbg.cnb.csic.es/PB/E1130	

In the main web form, simply paste your MSA and press "Next". The system also allows introducing un-aligned sequences, which will be aligned internally. The "Tools" section of the form contains an option for trying to identify the IDs of the sequences so that later they can be linked to the corresponding databases.

In the resulting page, you can see your alignment and color the residues according with different schemas with the menu above. The coloring affects only the residues above a conservation threshold, that can be changed in the accompanying menu. The "wrap" checkbox is similar to the "Wrap" option of Jalview and allows splitting the alignment in blocks to better visualize it. Sequences can be removed by dragging the IDs to the trash icon. That icon also allows recovering deleted items.

The "+" icon in the tools menu gives access to a more standard menu with more options. "Export" allows exporting the alignment to different plain text and graphical formats, including MS-Word. The "Java-based export" section of this menu allows directly opening web-based tools such as Jalview or phylogenetic tree viewers with the alignment preloaded (this requires JRE).

1.7.3 Summarizing MSAs

Sometimes, instead of showing the whole alignment, it is more convenient to generate graphical or textual representations which summarize the information contained on it. This is useful, for example, to create figures for publications or to communicate the main information contained in the alignment (conserved positions, motifs, etc.) without showing it as a whole.

Consensus Sequence

A consensus sequence for a multiple protein sequence alignment is a pseudo protein sequence which contains the most common amino acid at each position of the alignment, and especial symbols for the variable positions.

You can use ***EMBOSS cons*** web server to generate a consensus sequence from a multiple sequence alignment.

EMBOSS cons— Generation of consensus sequences of MSAs	http://emboss.bioinformatics.nl/cgi-bin/emboss/cons	
	https://csbg.cnb.csic.es/PB/E1140	

Just upload a MSA file or paste its contents (e.g. in FASTA format) and press "Run cons".

For example, for the following MSA:

```
PGGASK--VGQTDNDPQAIKDLP----PQGED------------------------
PGGAAK--VGKVANDPQFVKALP----QQAAHE-----------------------
PSRARRDAVG--DH--PAVEALP----PQSGPHKKEISFFTVRKEEAADADLWFPS
PGGSSN--VDLPIADSEYVKTLD----EK---------------------------
PGGAAK--LGRVGADTPLVAALPASTTPHP--------------------------
```

The consensus sequence calculated by EMBOSS_cons (with default parameters) is:

```
>EMBOSS_001
PGGASKxxVGKxxNDxQxVKALPxxxxPQxxxxxxxxxxxxxxxxxxxxxxxxxxx
```

You can also export the consensus sequence shown by Jalview: open the multiple sequence alignment then right-click on "Consensus" annotation row, and select "Copy consensus sequence". Now you can to paste the consensus sequence to your preferred text editor.

Sequence Logos

Sequence logos (Schneider and Stephens 1990) are graphical representations of the proportions of the different amino acids in the positions of multiple sequence alignments. For each position, a stack with the 20 amino acids is shown. The height of the letters is proportional to the frequencies of these amino acids at that particular position of the alignment. Apart from the graphical aspect, the main advantage respect consensus sequences is that sequence logos contain information not only on the most frequent amino acid, but also on others with also high frequency which could be informative. Additionally, since they are graphical representations, the letters can be colored according with different criteria. They are usually appropriate to

Fig. 1.4 Example of a MSA and its associated sequence logo generated by Weblogo

represent short motifs (consecutive positions), or individual positions (not consecutive) in an alignment (Fig. 1.4).

Weblogo (Crooks et al. 2004) is a web-based application that generate web logos for a multiple sequence alignment provided by the user.

Weblogo—Generation of sequence logos from multiple sequence alignments	weblogo.threeplusone.com/create.cgi	
	http://csbg.cnb.csic.es/PB/E1150	

To use it, simply paste or upload the multiple sequence alignment file, check default options and click the "Create Logo" button.

Most typical options you might want to change are:

- Output format: choose PNG or JPEG (for bitmap image formats) or EPS, PDF or SVG (for vector graphics formats).
- Color scheme: for protein sequences Hydrophobicity (the default), Chemistry, Charge or Custom to introduce your own.

1.8 Finding Similar Sequences in a Database (Advanced)

If you want to find remote homologues of your protein sequence you might need to use more sensitive methods than the basic BLAST search discussed in Sect. 1.6. Remote homologs are homologous proteins that diverged a long time ago and, consequently, they only share very subtle sequence signatures difficult to capture by a basic BLAST comparison. These more sensitive methods use profiles (see Sect. 1.8.1 bellow), and they look for matches between database sequences and a query profile, rather than sequence-sequence matches.

You should use these "advanced" methodologies if you have not found enough relatives of your query protein with a plain BLAST and/or you suspect that you are missing remote homologs. This could be especially important if you are looking for an homolog of known structure to model the 3D structure of your query protein (Sect. 2.5.1) and none of the "close" homologs found by BLAST is such.

For using these methodologies you have two main alternatives that will be extensively discussed below:

- Let the program build the profile for your query protein during an iterative search. You only need to provide the query sequence and decide how many iterations are needed.
- Build a "good" multiple sequence alignment by your own (previous section), and provide this MSA as input. The program will build a profile for that MSA and scan the databases for sequences or other profiles matching it.

1.8.1 Sequence Profiles

Sequence profiles are a widely used way of encoding the information contained in a multiple sequence alignment. Probably you will never construct a profile on purpose, but you will certainly use programs and databases that build profiles (even if you don't notice it). That is the reason we include here a very brief description on sequence profiles.

Sequence profiles are probabilistic models used to capture the characteristics of a multiple sequence alignment. As a probabilistic model, it is possible to evaluate to what extent a given protein sequence matches a profile, and give a numeric score for the match. Since they encode the characteristics of many proteins of the same family, they are better for finding new members of that family than a sequence comparison against a single representative. Profiles have been used to create most of the currently available family and domain databases (Sect. 1.9). They are also part of profile-based sequence search methods that have been proved to be highly sensitive in finding remote homologous proteins (see examples below).

Two common representations of sequence profiles are position-specific scoring matrices (PSSMs) (also known as weight matrices) and "hidden-markov models"

(HMMs). It is important to remember that a profile (either PSSM or HMM) is always built from a multiple sequence alignment. But the exact MSA used to build a profile cannot be recovered from its profile representation.

1.8.2 Iterative Profile Construction

Position-Specific Iterated BLAST (***PSI-BLAST***) (Altschul et al. 1997) is a program that searches in a database for similar protein sequences in an iterative way. The first iteration is a normal blastp search (Sect. 1.6.2). In the second and successive iterations, a profile (in the form of a PSSM) is built from the significant hits found in the previous iteration and used to look for new sequences matching it. You won't notice this, as it is internally handled by PSI-BLAST. You only need to let know PSI-BLAST whether you want to run a new iteration. Remember that you need to run at least two iterations to do an actual profile-sequence search.

In each iteration, PSI-BLAST marks out those new sequences found (that were not in the previous iteration) so that you can decide when to stop. You can stop, for example, if the search "converged" (no new proteins were found). Remember that, although the input to the program is a sequence, a profile is automatically built from the hits of the first iteration.

PSI-BLAST—Iterative BLASTp profile-based search	http://blast.ncbi.nlm.nih.gov/
	http://csbg.cnb.csic.es/PB/E1160

From NCBI's BLAST page, follow the "protein blast" link, select the "Database" to search against, and choose PSI-BLAST (Position-specific Iterated BLAST) in "Program selection". Then click the "BLAST" button. This will run the first PSI-BLAST iteration, returning a table with hits. Note that all alignments with E-value better than a given threshold are marked in the column "Select for PSI blast". Now you can proceed with the second iteration (click the "Go" button close to "Run PSI-Blast iteration 2"). In this second iteration PSI-BLAST builds a profile from the previous hits (note that, by default, a maximum of 500 hits will be used to build the profile), and searches for new sequences that match this profile. New sequences found in the current iteration are marked in yellow. Now you can repeat the process and run as many iterations as you will.

This is an interactive (besides iterative) process since it is difficult to automatically decide how many iterations to run. It depends on the particular protein/family you are working with, but in general there are two main reasons to stop. The most common one is that the search "converged" (no new sequences are found). It might also happen that the search "jumped" to an undesired region of the sequence space:

for example one or a few un-related proteins "leaked" into the profile and expands with new members in successive iterations.

1.8.3 HMM Profile Search Against a Sequence Database

The ***hmmsearch*** program within the HMMER suite (Finn et al. 2011) allows you to find sequences matching the multiple sequence alignment (MSA) you provide as input. As you might guess from its name, HMMER represents alignments (profiles) as HMMs. This is the appropriate program to use when you want to have full control on the MSA that is used in the search. In general, it is also more sensitive than PSI-BLAST since HMMs retain more information of the MSA than PSSMs and, consequently, they are able to pick up more remote homologs. Consequently, another reason for using this tool is when you want to capture even more remote homologs than with PSI-BLAST. The drawback is that you have to use a reliable MSA as input since the results totally depend on its quality.

HMMER hmmsearch— Search for sequences matching a MSA	http://hmmer.janelia.org/search/hmmsearch	
	http://csbg.cnb.csic.es/PB/E1170	

First of all you need to create a good multiple sequence alignment (MSA) that is representative of the family or domain that characterizes your protein sequence. For that, please refer to Sect. 1.7. You might also use a MSA already available in a database for your family (Sect. 1.9).

Once you have the MSA, paste it or upload the corresponding file (in FASTA format), select the sequence database to search, and optionally restrict by taxonomy. Finally click the "Submit" button.

The HMMER web site includes other types of searches that are not fully described here because they are similar in goal to other tools explained in this book:

- **phmmer** is equivalent to a blastp search.
- **jackhammer** is equivalent to PSI-BLAST. It uses an iterative approach similar to PSI-BLAST, but building HMMs instead of PSSMs.
- **hmmscan** is equivalent to HHPred, described in next point.

1.8.4 HMM Profile Search Against a Profile Database

HHPred (Soding et al. 2005) is a powerful tool to find remote homologues by comparing profiles (in this case represented as HMMs). Instead of comparing a profile against each sequence in a database (as done by hmmsearch), it compares a profile with a database of profiles. This means, in practical terms, that even more remote homologs can be found.

In general, our suggestion is to use HHPred when previous programs do not provide significant results. For example, HHPred is a good method to search for possible structural templates in PDB when no clear homologs are found by standard BLAST searches. This is because structurally similar proteins coming from a common ancestor share subtle sequence signatures which can be captured with these sensitive profile-based methods. See Sect. 2.5.

HHPred—Search for profiles matching a MSA	http://toolkit.tuebingen.mpg.de/hhpred	
	http://csbg.cnb.csic.es/PB/E1180	

Even if you can provide a single sequence as input to HHPred (in this case a multiple sequence alignment is automatically built for you), to fully control the search it is better to provide a MSA as input, and set the "Max. MSA Generation iterations" option of the web form to zero.

Then select the database of profiles to search against in "Select HMM databases" (multiple database selection is allowed). For example, if you want to look for templates for structure modeling select PDB as database. Then click the "Submit job" button, just above the "Search Options" section.

1.9 Protein Motifs, Domains and Families

There are a number of databases that contain information on known conserved protein families, motifs and domains. These databases are, in general, constructed using different approaches, so in some cases contain redundant information and in others complementary information. All of them are intended to classify protein sequences, and are very valuable resources to search for in order to characterize proteins by putting them in a functional, structural and evolutionary context.

A protein family is a group of related proteins. The grouping can be based on evolutionary, functional or structural reasons, although in general the agreement between them is high. In some cases, families can be hierarchically organized in superfamilies (e.g. Sect. 2.2.1). In many cases families are characterized by conserved sequence motifs which can be related to the function of that set of proteins.

A protein domain is a part of the protein sequence (or structure) that is conserved, and shows some sort of independence in terms of structure, function and/or evolution respect to the rest of the protein. Most eukaryotic proteins are multidomain. You can think of protein domains as the modular building blocks of proteins. Knowing the domain composition of a protein of interest is very useful not only for understanding its function, but also to design constructs/chimeras (to prove the individual function of parts of the protein or to express shorter versions of a long protein that does not crystallize).

The most convenient way to discover which families does your protein belong to, or which domains it contains is to search the InterPro with your protein sequence. ***InterPro*** (Hunter et al. 2012) is a resource that integrates several databases describing protein families, protein domains and protein motifs (among them Pfam, PROSITE, and SUPERFAMILY that will be explained later). When you perform a sequence-based search against the InterPro you are in fact performing several sequence-profile searches, against all the source databases it integrates.

InterPro—Integration of different family/domain databases	http://www.ebi.ac.uk/interpro	
	http://csbg.cnb.csic.es/PB/E1190	

From the InterPro home page, you can paste your protein sequence or ID and press "Search".

As an example, we will comment the analysis of the UniProt entry P08622, corresponding to the DnaJ protein of *Escherichia coli*. Enter P08622 in the Search form on top of the InterPro site, and press "Search". The result page includes an overall description of "Protein family membership" (Chaperone DnaJ IPR012724) and "Domains and repeats" displayed along the length of the sequence. You can see the details of the domain/repeat by pointing with the mouse over a certain colored region. E.g. the first domain extends from residue 1 to 106 and is labeled as "DnaJ domain (IPR001632)". All these correspond to the integrated view of families and domains provided by InterPro (and corresponding InterPro IDs, like IPR012724 or IPR001632).

Click on the family description ("Chaperone DnaJ") to obtain more information on the family, including relationships to other families, general description and Gene Ontology functional annotations. A menu on the left, labelled "Overview", gives access to the list of UniProt proteins belonging to that family ("Proteins matched"), the unique domain combinations of these proteins ("Domain organizations"), access to proteins by taxonomy ("Species") as well as other protein-related information (such as "Pathways&interactions", "Structures" and "Literature").

In addition to this summary information, a section with "Detailed signature matches" follows. For each family, domain, repeat or site matched, a list of signatures corresponding to the source databases integrated by InterPro is provided. For example, the N-terminal "DnaJ domain" (IPR01623) has been identified through the signatures (independently) established in Pfam (PF00226 DnaJ), Prosite (PS50076 DNAJ_2) and SUPERFAMILY (SSF46565) among others. You can further explore the information provided by the source databases by clicking on each of the links on the right.

Among the databases integrated in the InterPro resource we describe three of them in the following sections. They have been included to highlight different aspects of the information they provide and because they have particular functionalities not present in the integrated InterPro. For a complete description of the databases integrated in the InterPro refer to the InterPro release notes: "Member database information" (http://www.ebi.ac.uk/interpro/release_notes.html).

Pfam ("protein families") (Finn et al. 2014) is a database that contains information on conserved protein families and domains. These arise from curated alignments of proteins manually assigned to a family by expert knowledge ("seed" alignments) which are then expanded by matching new sequences against them in an automatic manner. Therefore, technically Pfam is in fact a collection of multiple sequence alignments, and relevant associated annotations. This is the way in which Pfam-A families are generated. Another section of Pfam (Pfam-B) contains a much larger collection of families generated automatically by clustering the sequence space. The alignments for both types of families are in fact represented in the computer as Hidden Markov Models (HMMs), statistical models that capture the essence of a protein family or domain (see Sect. 1.8.1). You don't need to know the details about HMMs to use Pfam, just be familiar with the term.

One of the utilities of Pfam is to obtain precompiled multiple sequence alignments for our protein of interest, which include distant homologs due to the sensitivity of the HMMs.

Pfam—Database of protein families and domains	http://pfam.xfam.org/	
	http://csbg.cnb.csic.es/PB/E1200	

You can do several searches on the Pfam web site, but most probably you will "arrive" to a Pfam entry once you have searched the InterPro, or by following a link from other protein sequence databases (such as the UniProtKB).

Each Pfam entry allows you to see its "Domain organization" (or set of unique architectures). I.e. all the combinations of this family/domain, with any other family/domain as found in the protein sequences stored in UniProtKB.

In "Alignments" you will find several multiple sequences alignments for that entry. These include the original seed alignment as well as the full alignment, alignments for representative proteome subsets, and those including sequences from the NCBI and metagenomics sequence databases. You can download or visualize these alignments, as well as a sequence logo ("HMM logo") of the seed alignment.

Finally, you can see the taxonomic distribution of the Pfam family/domain in the "Species" tab.

PROSITE (Sigrist et al. 2002) contains a collection of protein domains, families and functional sites. This collection is described in terms of sequence profiles and sequence patterns. For a description of sequence profiles refer to Sect. 1.8.1. What characterizes PROSITE, compared with other domain/family resources, is the presence of "sequence patterns". Sequence patterns represent usually short sequence motifs (around 10–20 amino acids) which are present in a number of proteins (that can be homologs or not). Within PROSITE, they are represented as a 'regular expression' which indicates which amino acids are allowed/disallowed in each position of the motif. The recurrence of these patterns in proteins is usually indicative of functional importance.

PROSITE—Database of protein families, domains and recurrent short sequence motifs	http://prosite.expasy.org/	
	http://csbg.cnb.csic.es/PB/E1210	

Individual positions of the sequence in a PROSITE pattern are separated by hyphens '-', and one-letter amino acid codes are used. Other special characters are:

- 'x' represents any amino acid
- '[]' lists all acceptable amino acids for a position
- '{ }' lists all amino acids that are not accepted in that position
- Repetition of an element is indicated with a number or a range between parentheses. E.g. x(2) corresponds to x-x
- '.' Designates the end of the pattern

For a more detailed explanation of PROSITE patterns and profiles refer to PROSITE user manual (http://prosite.expasy.org/prosuser.html).

For example, the PROSITE pattern describing the DNAJ_1 motif (PS00636) found in the *E. coli* DnaJ protein sequence (see http://prosite.expasy.org/PS00636) is expressed as:

```
[FY]-{GL}-x-[LIVMA]-{IP}-x(2)-[FYWHNT]-[DENQSA]-x-L-x-[DN]-x(3)-[KR]-
{F}-{P}-[FYI].
```

This pattern represents a sequence motif of 20 residues. In the first position, either F or Y can be found. The second allows any amino acid except G and L. The third (x) allows any amino acid, etc…

SUPERFAMILY (de Lima Morais et al. 2011) is a collection of sequence profiles (in the form of HMMs) for the groups of homologous domains ("superfamilies") of known three dimensional structures defined in the SCOP database (see Sect. 2.2.1). Therefore, a matching of a sequence against a SUPERFAMILY entry allows not only to infer its structure, but to put it into SCOP's hierarchy, with all the evolutionary and functional implications it has.

SUPERFAMILY—HMM profiles for SCOP superfamilies	http://supfam.org/SUPERFAMILY/	
	http://csbg.cnb.csic.es/PB/E1220	

The contents of the database can be browsed according with different criteria ("Browse" section on the left), or searched by keyword or sequence similarity. The last is the most useful way of accessing SUPERFAMILY since, as explained above, a match of our query sequence against a SUPERFAMILY entry provides important structural and functional information.

1.10 Basic Phylogeny

A phylogenetic analysis of your protein sequence of interest and its relatives allows obtaining important information on the evolutionary history of that family of proteins. The main result of such analysis is usually a phylogenetic tree of the family, which is a graphical representation of the evolutionary relationships among its members and can be used, for example, to reconstruct ancestral sequences, to study the eventual co-evolution of this family with others, and, in general, to obtain knowledge on how this family originated and evolves.

Although carrying out a thorough phylogenetic analysis is not trivial and requires sophisticated tools as well as a great deal of expert knowledge, some basic analysis can be performed with the kind of web-based tools addressed in this book.

The first step for performing a phylogenetic analysis is to retrieve homologous sequences of our protein of interest, from the same and/or different species, and to multiply-align them (see Sects. 1.6 and 1.7). Most servers for performing sequence searches (Sect. 1.6) have, within the same interface, an option to "pipe" the results of a search (list of homologous) to a multiple sequence alignment and/or phylogenetic analysis tool. For example, in the BLAST server at the NCBI (Sect. 1.6.2), this can be done with the "Multiple alignment" link at the "Descriptions" sections of the result page, after selecting the hits we want to align.

In case you already have a multiple sequence alignment of our family of interest (Section 1.7) there are a number of tools aimed at generating (and visualizing) a phylogenetic tree from it.

The ***ClustalW2 phylogeny*** tool at the EBI generates and displays a phylogenetic tree from a user-provided alignment using two tree-generation methods implemented in ClustalW (Chenna et al. 2003): neighbor joining (NJ) and UPGMA. While these methods are not the state-of-the-art in phylogeny, they are adequate for a simple and fast phylogenetic reconstruction.

```
(
(
e-coli:0.20425,
y-pestis:0.22683)
:0.02449,
(
s-typhimurium:0.28355,
(
c-crescentus:0.49997,
p-multocida:0.31285)
:0.05135)
:0.02242,
v-cholerae:0.25173);
```

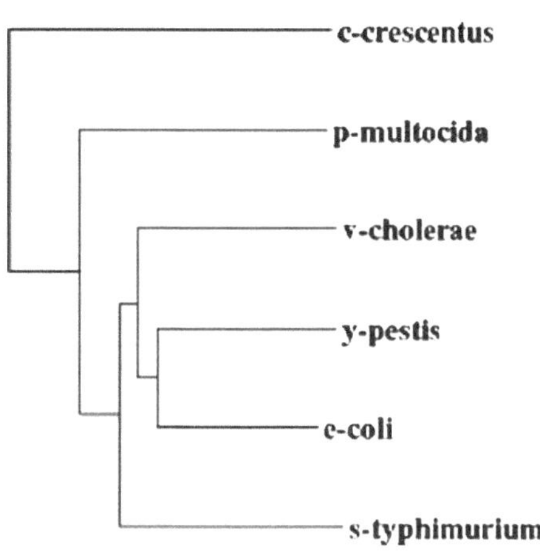

Fig. 1.5 Example of a phylogenetic tree with 6 sequences (right) and its encoding in a newick file

ClustalW2 phylogeny—Generation of phylogenetic trees from multiple sequence alignments	http://www.ebi.ac.uk/Tools/phylogeny/clustalw2_phylogeny/
	http://csbg.cnb.csic.es/PB/E1230

In most cases, the default parameters of the web form are adequate and you have just to provide the input multiple sequence alignment and press "Submit". The results page contains the resulting tree in "newick" format and a simple graphical representation at the bottom. In the newick format, the hierarchical phylogenetic relationships among proteins (leaves) and internal nodes (ancestral states) are represented by nested parenthesis (Fig. 1.5). The "View phylogenetic tree file" option can be used to retrieve and save the tree (in newick format) for further processing or for sending it to more advanced tree visualization tools (see next point).

The same tree can be graphically represented in different ways depending on the features you want to highlight or simply due to aesthetic reasons. The ***Phylodendron server*** at the Indiana University takes as input a tree in newick format and generates different graphical representations for it.

Phylodendron—Graphical representation of phylogenetic trees	http://iubio.bio.indiana.edu/treeapp/treeprint-form.html
	http://csbg.cnb.csic.es/PB/E1240

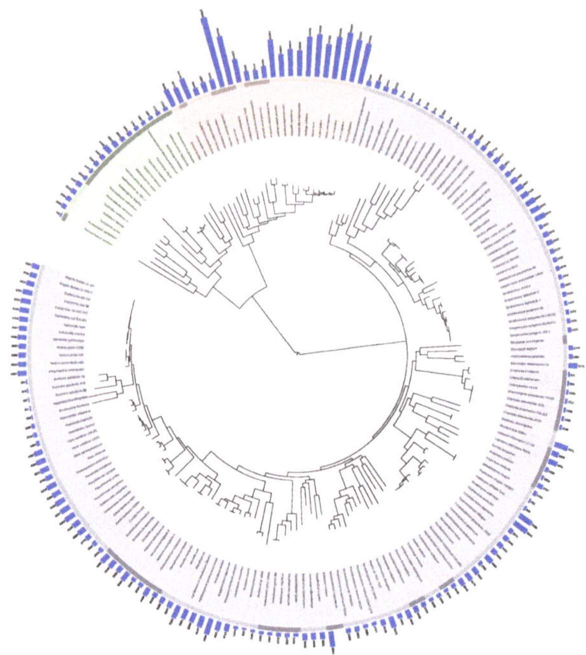

Fig. 1.6 iTol circular representation of a tree with two datasets associated to the nodes: a qualitative one (represented by the background color) and a quantitative one (blue bars)

Select the tree "style" for the graphical representation at the top of the web form, and the characteristics of the final image (format, size in pixels, text fonts, …) in the "Output" section.

Although very simple to use, and adequate for a quick visualization of our tree, the capabilities of the Phylodendron server are quite limited, and it generates static representations (image files) of the trees. A much more advanced tree viewer is the ***iTol server*** at the EMBL. iTol (Letunic and Bork 2006) generates different interactive representations of phylogenetic trees. iTool allows associating generic data (textual, numeric …) to the tree leaves/nodes, so that these can be shown in the tree representation (Fig. 1.6). The interactivity of iTol also allows working and visualizing large trees, since the user can zoom in and move to a region of interest. Moreover, the iTool server provides access to pre-generated trees, such as a "Tree of Live" (tree representing the evolutionary distances between known species) so that you can map your datasets on them.

iTol—Graphical interactive representation of phylogenetic trees and datasets associated to their elements	http://itol.embl.de http://csbg.cnb.csic.es/PB/E1250	

The main entry point for visualizing our tree and associated data in iTool is the "Data upload" tab at the top. In the "Upload your own tree" section of the web form, you can specify the file containing your tree (newick and other formats are accepted). This is the only input required. Pressing "Upload" takes you to a page with an internal ID you can save for retrieving the tree later. Within this page, follow the "go to the main display page" link to visualize the tree (see below).

To annotate trees use the "Upload datasets for your tree" section. You can upload up to 10 files containing attributes and data associated to the elements of the tree (leaves –proteins- and internal nodes). The format of these files varies slightly depending on the nature of the data we are importing, but in the simplest case contains just a list of pairs "node <SEP> value", where the separator "<SEP>" is specified with the "field delimiter" option of the form. For example, if you have a tree relating 5 proteins (protein1…protein5) and want to display in the tree bars proportional to their molecular weights, you can specify those in an (comma-delimited) file such as:

```
protein1,140
protein2,100
protein3,230
protein4,220
protein5,115
```

Select from the "Data type" menu the type of representation you want for your data. The exact format of the file for each type of data is extensively explained in the "help pages" link (section "uploading datasets"), including examples. These examples are very clear and taking a look at them is the best way of getting familiar with these dataset files. When colors have to be specified within this file, these have to be coded as "web hexadecimal RGB" strings (e.g. "#ff0000" for red, "#A15D2A" for brown, …). There are many web sites with color browsers and tables where you can obtain these codes for a given color or vice versa, for example: www.rapidtables.com/web/color/RGB_Color.htm.

A number of additional fields might show up in this form depending o the "Data type" selected (e.g. circle radius for a pie chart, bar size for a simple bar chart, etc). Finally, if you are importing more than one dataset (files) you can associate different colors to them with the "Dataset X color" field. Note that you can also use the color browser within this field to obtain the RGB strings for your colors of interest (previous paragraph).

To retrieve a previously uploaded three, introduce its ID in "Retrieve a previously uploaded tree".

The tree display page contains an interactive representation of the tree that can be zoomed in/out with the "+" and "-" buttons on the left. The "Search" link close to them allows searching for a given leaf (protein). The tree can be moved through the canvas by dragging it with the left mouse button. The type of representation for the tree (circular, cladogram, …) can be changed with the "Mode" option of the "Basic controls" tab. For most changes in the representation to take effect, it is necessary to press the "Update tree" button on the left. The "Advanced controls" and "Display controls" tabs can be used to change other features of the representations. Their items vary slightly depending on the characteristics of our tree (e.g. an option may exist for showing the bootstrap values of the internal nodes in case these are present). Of special interest is the option to change the size of the fonts used to display, for example, the leaf names. In case we have uploaded datasets for our tree (see above), additional controls might appear in the canvas, for example to show/hide the representation(s) of these datasets. Finally the "Export tree" button on the left takes us to a form where you can export your representation (tree and eventually associated datasets) to a number of graphical formats.

The tools commented so far cover the most basic tree generation and visualization operations. Users requiring more advanced features, such as state-of-the-art tree generation methods, advanced tree viewers, etc. might try the ***Phylemon2 server*** at the CIPF (Sanchez et al. 2011). Although there is the possibility for the user to open a personal account in order to better control the submitted jobs, etc., any user can start using the server right away through the "start as anonymous user" option on the right.

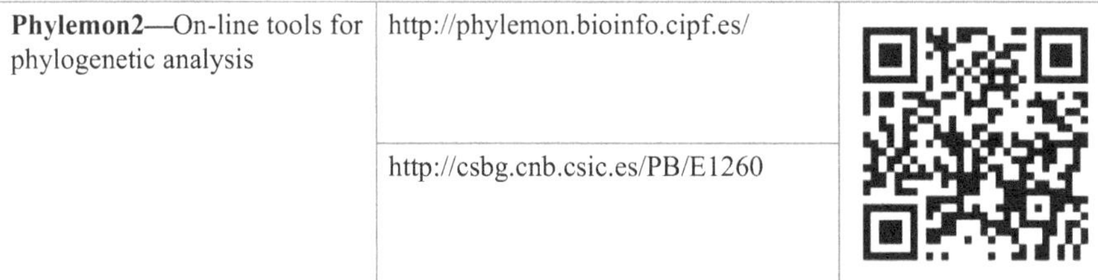

Phylemon2—On-line tools for phylogenetic analysis	http://phylemon.bioinfo.cipf.es/	
	http://csbg.cnb.csic.es/PB/E1260	

The "Phylogeny" tab at the top of the page is the entry point to access the phylogenetic tools interfaced in this server. Each tool takes to a web form where you can specify the input file and a number of parameters, as well as a name/description for the job (see below). Follow the "Help" link to obtain information on that specific tool, its input, parameters, etc. There is also the option to automatically fill the form with example data. Since advanced phylogenetic algorithms can take some time to run, all the operations are performed under a job control system. Submitted jobs are queued and run when resources are available. The column on the right contains information on the jobs submitted and their state ("waiting in queue", "running", "ready", etc.). The list is automatically updated and the results of a job can be accessed once it is finished (green). The same column also allows accessing the datasets (e.g. input files) associated to the jobs. When using the system as "anonymous", all jobs and associated data are deleted in 24 h. The results of a job include a number of "Send to .... tool" options so that these can become the input of other tools within the system (for example the tree generated can be send to a tree viewer).

Within the many options available in Phylemon, we are going to mention a tree construction method not available in ClustalW2: maximum likelihood (ML). ML is available under the "Phy-ML" tool of the "Phylogeny" tab. The input for this tool is a set of unaligned sequences in Phylip format. Take a look at the provided example and Sect. 1.7.1 for information on format conversion. Phylemon also offers a format converter ("File format conversion" in the "Utilities" tab). In "Data type" select "amino acid". Once the job is finished, the results page contains a link to the tree file in newick format ("PhyML Tree—view") and an option to visualize the tree with two viewers included in the system ("Send to .... tool").

These are interactive viewers where you can manipulate and change the layout and characteristics of the tree representations on the fly. These tree viewers can also be used with trees generated outside Phylemon. They are available in the "Viewers" section of the "Utilities" tab and take as input trees in newick format.

Another web server which gives access to a large number of tools related to sequence search, analysis and phylogeny through simple interfaces is ***Phylogeny.fr*** (Dereeper et al. 2008).

Phylogeny.fr—Online tools for phylogenetic analysis	http://www.phylogeny.fr/	
	http://csbg.cnb.csic.es/PB/E1270	

The "Phylogeny analysis" tab gives access to predefined complete workflows for phylogenetic analysis, e.g. "multiple sequence alignment>alignment refinement>tree generation>tree visualization". The individual tools are available under the "Online programs" tab. As an example, we are going to comment here the tool for generating Bayesian trees so as to cover the three main tree generation strategies: in increasing order of complexity/accuracy: neighbor joining (ClustalW2), maximum likelihood (Phylemon) and Bayesian. This tool is available at "Online tools>Phylogeny>MrBayes". Select "protein" as data type. Then, browse to the file containing the multiple sequence alignment (different formats are accepted, including FASTA). Since the default parameters involve relatively long runs, you have to provide an email address for being informed of job completion. The results page contains a link to the tree file (newick format) as well as other MrBayes results, including an approximate representation of the tree in text format. It also contains a link to interactively visualize the tree ("Visualize the tree with ATV"), an operation which requires Java to be installed in your system.

Chapter 2
Structures

2.1 Introduction

The three-dimensional (3D) structure of a protein contains a lot of information on its function, and can be used for devising ways of modifying it (propose mutants, protein design, etc.).

This chapter covers some of the bioinformatics technologies aimed at managing, storing and making computations on protein 3D structures. We will start by getting familiar with the on-line resources where primary and derived information on protein structures is stored, so that we can search and retrieve the available structural information for our protein(s) of interest. Then we will learn the basis on how to manipulate visualize and compare protein 3D structures. Since the experimental structural information is limited to a relatively small number of proteins, we will learn how to predict structural features for raw protein sequences, from low level features such as secondary structure, coiled-coil regions, etc. to complete 3D structures for protein chains. Finally we will take a look at some bioinformatics methods aimed at extracting useful information from these 3D structures (either experimental or predicted).

2.1.1 Storing Protein Structures—The PDB File Format

In the same way protein sequences are stored in text files with particular formats so that they can be interchanged between different programs, there is a widely used format for computationally store protein 3D structures: the "protein data bank" (PDB) format. Almost all programs and web servers working with protein structures are able to handle PDB files. In most cases you will manage PDB files (e.g. for interchanging structural information between diverse servers) without caring about their content or format. Nevertheless, it may be useful to be familiar with their format since some operations with protein structures can be performed by directly editing these files.

F. Pazos, M. Chagoyen, *Practical Protein Bioinformatics*,
DOI 10.1007/978-3-319-12727-9_2

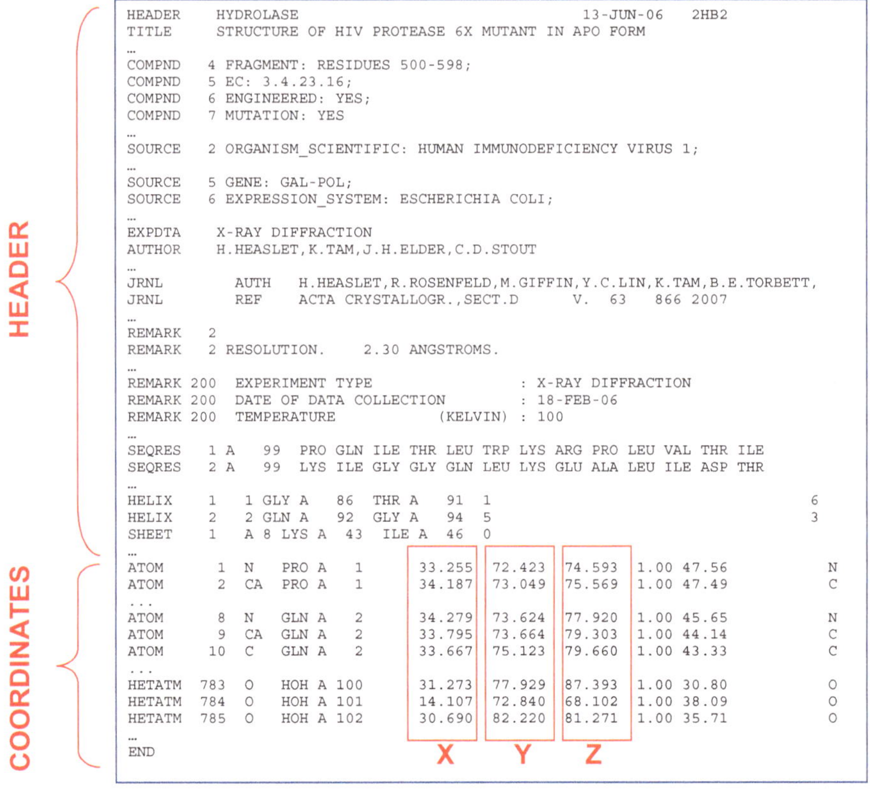

```
HEADER    HYDROLASE                               13-JUN-06   2HB2
TITLE     STRUCTURE OF HIV PROTEASE 6X MUTANT IN APO FORM
...
COMPND    4 FRAGMENT: RESIDUES 500-598;
COMPND    5 EC: 3.4.23.16;
COMPND    6 ENGINEERED: YES;
COMPND    7 MUTATION: YES
...
SOURCE    2 ORGANISM_SCIENTIFIC: HUMAN IMMUNODEFICIENCY VIRUS 1;
...
SOURCE    5 GENE: GAL-POL;
SOURCE    6 EXPRESSION_SYSTEM: ESCHERICHIA COLI;
...
EXPDTA    X-RAY DIFFRACTION
AUTHOR    H.HEASLET,K.TAM,J.H.ELDER,C.D.STOUT
...
JRNL        AUTH   H.HEASLET,R.ROSENFELD,M.GIFFIN,Y.C.LIN,K.TAM,B.E.TORBETT,
JRNL        REF    ACTA CRYSTALLOGR.,SECT.D      V.  63   866 2007
...
REMARK    2
REMARK    2 RESOLUTION.    2.30 ANGSTROMS.
...
REMARK 200  EXPERIMENT TYPE                : X-RAY DIFFRACTION
REMARK 200  DATE OF DATA COLLECTION        : 18-FEB-06
REMARK 200  TEMPERATURE           (KELVIN) : 100
...
SEQRES   1 A   99  PRO GLN ILE THR LEU TRP LYS ARG PRO LEU VAL THR ILE
SEQRES   2 A   99  LYS ILE GLY GLY GLN LEU LYS GLU ALA LEU ILE ASP THR
...
HELIX    1   1 GLY A   86  THR A   91  1                                   6
HELIX    2   2 GLN A   92  GLY A   94  5                                   3
SHEET    1   A 8 LYS A   43   ILE A   46  0
...
ATOM      1  N   PRO A   1      33.255  72.423  74.593  1.00 47.56           N
ATOM      2  CA  PRO A   1      34.187  73.049  75.569  1.00 47.49           C
...
ATOM      8  N   GLN A   2      34.279  73.624  77.920  1.00 45.65           N
ATOM      9  CA  GLN A   2      33.795  73.664  79.303  1.00 44.14           C
ATOM     10  C   GLN A   2      33.667  75.123  79.660  1.00 43.33           C
...
HETATM  783  O   HOH A 100      31.273  77.929  87.393  1.00 30.80           O
HETATM  784  O   HOH A 101      14.107  72.840  68.102  1.00 38.09           O
HETATM  785  O   HOH A 102      30.690  82.220  81.271  1.00 35.71           O
...
END
```


Fig. 2.1 Representative sections of a PDB file for storing information on macromolecular structures

A PDB file is a plain-text file in which the information related to a macromolecular structure involving one or more proteins is stored. Basically, it contains a header section with the "metadata" associated to that structure (protein name, experimental method used for structure determination, …) followed by the Cartesian coordinates (X, Y, and Z) of the atoms (Fig. 2.1).

The header section contains information on the biomolecules whose structures are represented in the file (name, organism, mutations, amino acid sequences, IDs in sequence databases…), the experimental/computational method used for determining/predicting it (e.g. X-ray crystallography, NMR, computational modeling, …) as well as the parameters and experimental details associated to these methodologies. This section also contains the bibliographic reference of the structure determination and information on non-protein molecules present in the structure (cofactors, ions, water molecules …). It may also contain information on the secondary structure content and the binding/functional residues annotated by the generators of the structure.

In the coordinates section (lines starting with "ATOM"), each atom is represented in a line which contain the atom number, its type within the protein (e.g. N: backbone nitrogen; CA: alpha carbon; CB: beta carbon…), the residue type it belongs to and its number within the sequence, the chain (for files containing various protein polymers –complexes, multimers, …-) and the X, Y and Z coordinates in angstrom (Å). The rest of the line contains variable information depending on the method: e.g. B-factor for X-ray structures, confidence figures for predicted models, etc. The "HETATM" lines contain the coordinates of the atoms of non-protein molecules, such as prostetic groups and waters (Fig. 2.1).

If the PDB file contains a number of different alternative structures for the same protein (e.g. alternative computational models or an ensemble of structures compatible with NMR data) these are stored in different "MODEL" sections.

Editing this file it is possible, for example, to extract a chain of interest from a file with multiple chains: simply copying/pasting the ATOM and HETATM lines for that chain to another file. Similarly, by copying/pasting we can extract a representative model from a PDB file representing an ensemble of alternative structures (i.e. NMR). It is also possible to manually remove solvent/water molecules or the atom lines of the residues which are causing problems in some programs.

2.2 Main Protein Structure Databases

There are many on-line databases with primary and derived information on protein three-dimensional (3D) structures.

The original database in which primary information on protein 3D structures is deposited is the ***Protein Data Bank (PDB)*** at the RCSB (Berman et al. 2000).

PDB—Main primary database on protein 3D structures	http://www.rcsb.org http://csbg.cnb.csic.es/PB/T1010	

Apart from hosting the raw data on protein structures, a number of browsing, searching and analysis features were incorporated to the PDB with the time. For a molecular biologist, the main entry points to this massive amount of structural information are the "Search" and "Explore archive" sections of the web interface.

The search form, at the top of the main page, contains a single text entry box for quick searches based on the name of the protein, its sequence, the ligand co-crystallized, etc. Examples are provided when selecting these different search criteria. A more advanced search form is available at "Advanced search". This allows creating complex queries combining an arbitrary number of criteria. For that, select a search criterion in the "Choose a query type" selector and enter the search value(s) in the corresponding box (for example "structure title"/"hemoglobin"). Then add extra search panels ("+" button) and fill them with additional pairs search criterion/value (e.g. "chemical ID"/"HEM" and "X-ray resolution"/"between 0.0 and 2.0"). The combination of these three search criteria results in the complex query "structures of hemoglobins with resolution 2.0 Å or better and bound to heme groups". Within each panel, the "Result Count" button shows how many entries fulfill that particular search criterion, independently of the others. Once the complete query is constructed, press "Submit query". Close to that button, you have the option of filtering the results by sequence identity, so as to avoid retrieving, for example, many times the same protein crystallized with different cofactors. The results page lists the entries fulfilling all the search criteria together. At the top of the results page, a "filter refinements" panel allows further filtering the results (by organisms, etc.)

A particularly useful search criterion is "sequence (BLAST/FASTA/PSI-BLAST)". This allows looking for proteins of known structure (PDB entries) similar to a sequence of interest. This is the more direct way to check whether our protein has been crystallized, either itself or a close homolog that would allow building a 3D model by homology (Sect. 2.5.1). In the panel associated to that search criterion, paste the sequence of your protein and select the sequence search method as well as the cutoffs of minimum sequence identity or maximum E-value (see Sect. 1.6.2.1). Note that this can also be done in most BLAST servers selecting "PDB" or "RCSB" as the database to search against. The advantage of using this option within the RCSB search page is that the result of the sequence search can be, in a single shot, filtered by additional structural criteria adding more panels as explained above. This allows, for example, looking for homologous of known structure of our protein of interest crystallized with good resolution in a particular model organism.

With the "Explore archive" panel, you can successively narrow the scope of the search click by click until you reach the desired entry(ies). For the example of the hemoglobins above: release date: 2010-today (~38,000 entries) > organism: homo sapiens (~10,000 entries) > experimental method: solution NMR (~700 entries) > enzyme classification: isomerases (11 entries).

In all these cases, you finally end up in a list of entries of interest. In this list, for each entry you can see its PDB ID. This 4-character code (being the 1st character always a digit) is the main identifier of the entry and the standard way of referencing it. There is also a small picture of the structure and some basic information of the entry (title, release date, bibliographic reference, …). Just below the PDB code, there is a link to download the raw PDB file with the whole entry in text format. We will need this file if we want to further use the 3D structure outside the RCSB. Another link below the PDB code ("3D view") opens an interactive 3D viewer (JMol) where the structure can be inspected (rotated, zoomed, …) (Sect. 2.3.1). Finally, clicking the title of the entry, the page with all the information associated to that entry appears.

This information is split in different sections (tabs at the top). Within the "Summary" tab, we find additional images of the structure on the right, including the (eventually different) asymmetric unit of the crystal and the predicted biological

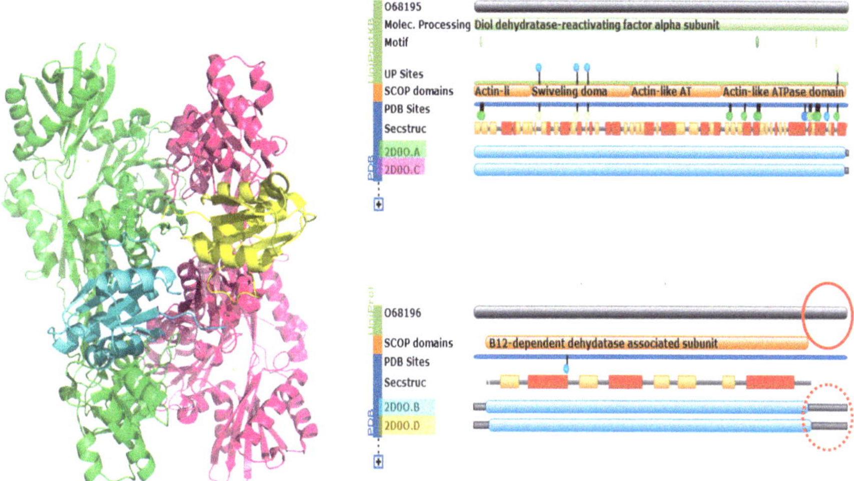

Fig. 2.2 Three-dimensional structure and RCSB molecular description for PDB entry 2D0O. This entry contains 4 protein chains: two dimers of the large and small subunits of the diol dehydratase-reactivating factor. Chains A and C (*green* and *pink*) are the large subunits, and B and D (*blue* and *yellow*) the small ones. Consequently, there are two different proteins within this entry (Uniprot IDs O68195 and O68196-right-). The molecular description panel also shows that the C-terminal part of the small subunits (O68196, chains B and D) has not been crystallized (*red circles*)

assembly. This panel also contains links to visualize/manipulate the structure in a number of structure viewers (Sect. 2.3.1). Another interesting panel of the "Summary" tab is "Molecular description", which contains a linear representation of the protein chains represented by that entry (together with their accessions in Uniprot) indicating which segments are actually in the PDB file (i.e. for which the 3D structure was determined), together with additional features for the proteins (active sites, domains, secondary structure, etc). This provides a quick overview on the global functional and structural characteristics of the proteins, as well as on the parts of the protein sequence which are actually crystallized (e.g. the structure could have been determined for a segment of the protein only) (Fig. 2.2).

The "Sequence" tab displays more detailed information on the sequence of the protein chains, together with their secondary structures and the residues involved in binding ligands. Finally, the "Links" tab contains links to the pages for that entry in other structure-related databases, mainly secondary databases based on PDB.

One of such PDB-derived databases is the ***PDBsum*** at the EBI (Laskowski et al. 2005). The original goal of PDBsum is to "enrich" the raw structural information of the PDB adding representations and data extracted and/or calculated from it. Although, as commented above, the current version of PDB also includes a lot of derived data, there are still a number interesting and unique features in PDBsum.

PDBsum—Derived data on protein 3D structures	http://www.ebi.ac.uk/pdbsum/ http://csbg.cnb.csic.es/PB/T1020	

PDBsum's top page allows to search by the 4-character PDB code, and by text using other information associated to the PDB entries and the protein chains within them (protein name, author, IDs in different sequence databases, …). It is also possible to look for entries similar to a given sequence provided by the user, as explained earlier for PDB.

The main page for a given entry shows on the left a picture of the structure and a summary of the molecules within it (protein chains, DNA/RNA chains, heteroatoms –ligands-, metals, waters, etc.). Three orthogonal views of the structure are available with the "eye" icons, and a link to interactively inspect that structure in a JMol viewer is also provided (see Sect. 2.3.1). On the middle column, you find a summary of the fields within the header of the PDB file (title, molecule name, author, …) and a "Links" section with hyperlinks to the corresponding pages for that PDB code in other structural databases). Here you also find a linear representation of the protein chains within that entry highlighting their coverage respect to the whole-length protein deposited in the sequence database, the annotated domains, functional sites, secondary structure, etc., similar to the "molecular description panel" of the RCSB commented above. On the right, we find the Ramachandran plot of the structure, which is a link to a complete report of Procheck (Laskowski et al. 1996), a program for evaluating the quality of protein structures. This top page also shows the abstract of the publication describing the determination of the structure and selected figures from that article, as well as a list of other papers citing it.

The "Protein" tab contains detailed information, at the residue level, of the protein chains within the entry (conservation, secondary structure, etc.). Here, the protein chains are also split in domains according with the CATH database (Sect. 2.2.1) and representations of the topology of these domains are shown on the right.

The "DNA/RNA", "Ligand" and "Metal" tabs contain diagrams showing the interactions between these biological entities and the protein chains in the structure (type of interaction, protein residues involved, …). The diagrams are generated with the LIGPLOT software (Wallace et al. 1995). For multi-chain PDB entries, the "Prot-prot" tab contains information on the interaction surfaces (interfaces): size, residues involved, … Similarly, the "Clefts" and "Tunnels" tabs contain the elements of this kind detected in the structure. These elements are related to binding and functional sites in many cases and, in combination with other evidences (clustering of residue conservation, etc.) can help in the prediction of these important regions.

2.2.1 Classifications of Structural Domains

Both PDB and PDBsum store protein structures not framed in any classification schema. Some PDB-derived databases try to classify the PDB entries in a hierarchical schema based on their structures, sequences, and inferred evolutionary relationships. Knowing the position of our protein of interest in these classifications might provide important information on its evolution, function and relatives.

Since domains are the structural, functional and evolutionary units of proteins, the protein structure classification schemas we are going to discuss below have the domain as the basic unit, and not the protein chain or the PDB entry (which can contain multiple chains eventually of different proteins). For example, the PDB entry 2d0o (Fig. 2.2) contains 4 protein chains: two hetero-dimers of the large and small subunits of the diol dehydratase reactivating factor. The small subunits (two chains within this entry) are mono-domain chains, while the large subunits (the other two chains) are chains with two different domains. Consequently, there are three different protein domains within this PDB entry, and indeed it is associated to three entries in the structural classification databases.

One of such databases is the ***Structural Classification of Proteins (SCOP)*** at the MRC (Andreeva et al. 2004).

SCOP—Hierarchical classification of protein domains of known 3D structure	http://scop.mrc-lmb.cam.ac.uk/scop/index.html http://csbg.cnb.csic.es/PB/T1030	

SCOP classifies protein domains according with a hierarchical schema in which at the top level we find the "structural classes". A given structural class is divided into "folds", which in turn contain "superfamilies". Superfamilies can be split into "families" and, finally, these contain the "domains" (Fig. 2.3).

Structural classes group domains according with their global composition of secondary structure elements (α-helix, β-strand, turns, etc.) Consequently, a class contains all the domains comprising α-helices only, other those comprising a mixture of α-helices and β-strands, etc. Some classes may also reflect broad aspects of the arrangement of the secondary structure elements. Classes are clearly artificial since, in general, they do not respond to any evolutionary or functional criteria. Moreover, in an attempt to comprehensively include all proteins available in PDB, this level comprises even more artificial classes defined based on methodological aspects, such as "low resolution structures", "small peptides", etc.

A structural class is subdivided into "folds". A fold groups those domains with the same content and 3D arrangement of secondary structure elements. For example, within the "all-α" structural class we find many different folds depending on the number and relative orientations of the α-helices. Folds are not homogeneous

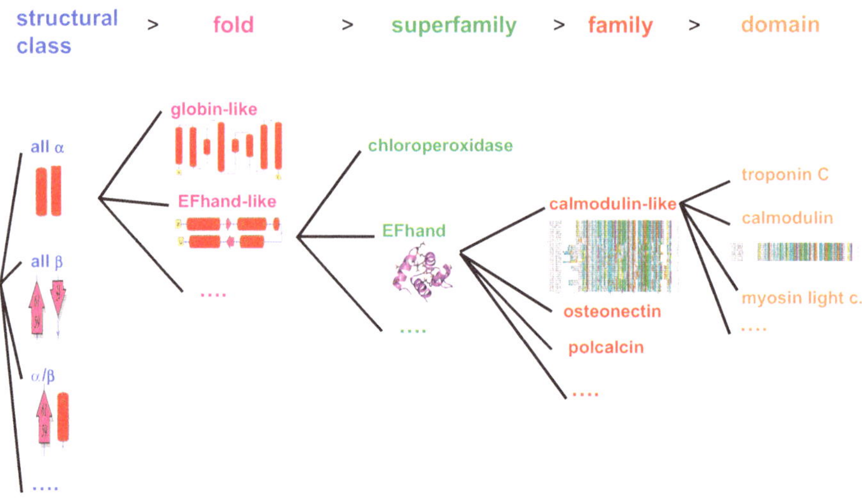

Fig. 2.3 SCOP hierarchical schema for classifying protein domains

in evolutionary or functional terms, in general. A given fold may comprise proteins with very different evolutionary origins and functions. However, it is at this level where the classification starts having some biological rationale since folds represent stable 3D layouts of protein chains repeatedly used by evolution due to their stability and functional advantages.

Within a given fold, superfamilies group homologous domains, that is, domains arisen from a common ancestor. The assignment of homology may be based on a clear sequence identity between the two domains or on subtler evidences, such as short sequence and structural motifs, and/or expert knowledge. For example, the "spectrin repeat-like" fold (all-α class) contains 16 different superfamilies, which are evolutionary unrelated even if they all have the same global 3D structure (number and arrangement of the α-helices). Domains within the same superfamily also share functional aspects, retained from the ancestor they all come from. We could say that the superfamily is the most interesting level from a practical point of view, since functional similarities (even if subtle) start to be forged here. Regarding structural similarities, even if they start at the fold level, in practical terms predicting fold in absence of any sequence signal is very difficult (Sect. 2.5). Consequently, the superfamily provides the best context for inferring structural and functional clues about our protein of interest.

Going deeper in the hierarchy, families group proteins with a clear sequence relationship. The functional similarity between members of a family is higher than between members of a superfamily. While both, families and superfamilies, reflect homology relationships (common ancestry), usually these are referred to as "close homology" and "remote homology" respectively. Finally, a given family contains the individual protein domains (the domain in different organisms (orthologs), mutants and other variations of the same sequence in PDB, etc.), and can eventually be be further "subdivided" into "subfamilies".

SCOP's web interface is very simple and the two main entry points to this resource are the "Keyword search" and "Enter SCOP at the top of the hierarchy" options, which are in the "Access methods" section of the home page.

Entering SCOP at the top of its hierarchy shows a list with the structural classes (first level in SCOP's hierarchy). Clicking a class expands it in the folds it contains, which in turn can be expanded to superfamilies, and so on up to the individual domains. For all the levels there are links to interactively visualize the corresponding domain (or a representative of the fold, superfamily, …). For the superfamilies, there are links to the corresponding entry in the Superfamily resources (section 1.9).

For users not familiar with SCOP's internal IDs and keywords, the "Search" facility is mainly limited to PDB IDs. As commented above, searching for a given PDB ID might result in several SCOP entries depending on the number of individual domains represented by that particular structure. If we want to locate a sequence of interest (or a close relative) in SCOP, it is better to look for it in PDB or other resources which provide much better search features (including by sequence-similarity) and then follow the outgoing links to SCOP.

The ***CATH*** resource at the UCL (Pearl et al. 2005) is also intended to classifying all known protein domains in a hierarchical structure. There are some conceptual and implementation differences with SCOP and, consequently, these two resources complement each other.

CATH—Hierarchical classification and annotation of protein domains	http://www.cathdb.info/ http://csbg.cnb.csic.es/PB/T1040	

CATH's hierarchy is slightly different to SCOP's, and its levels are "class", "architechture", "topology" and "homologous superfamily", followed by a number of "subfamily" levels defined based on arbitrary thresholds of sequence identity. The "topology" and "homologous superfamily" levels are equivalent to the "fold" and "superfamily" levels of SCOP. Another difference is that CATH has a strong emphasis in annotating raw genomic sequences based on matching against the profiles of its superfamilies (implicitly annotating also at the topology and upper levels). Consequently, CATH extends beyond proteins of known structure and includes all raw sequences that can be matched against its superfamilies. SCOP relies on external resources (SUPERFAMILY, Sect. 1.9) for that. Another characteristic of CATH is the integration of functional information from other resources (EC number, GeneOntology annotations, …) into its entries.

As in SCOP, the main entry points to CATH are the "Browse" and "Search" options, at the top of its main page. The Browse option allows navigating CATH hierarchy by expanding/collapsing nodes within the different levels. During this navigation, a panel on the left displays the information of the expanded node. The

"Search" form allows looking for CATH entries based on a number of different IDs (including CATH internal IDs, PDB IDs, etc.) as well as on keywords. There is also the possibility of directly searching by sequence similarity. Another interesting possibility is to search for structural similarity ("Search by PDB structure"). The user uploads a protein structure in PDB format and the system looks for CATH entries with similar 3D structure (see Sect. 2.3.2.2).

2.3 Structure Manipulation, Visualization and Comparison

In the previous section, we have listed some resources which can be used to look for the available structural information of your protein of interest and its relatives, as well as to put it in the context of a protein structural/evolutionary classification. In that section we also tangentially touched the issues of visualizing, manipulating and, to a lesser extent, comparing protein structures. All these will be expanded in this section.

Note that most tools commented in this section, at least those that admit a generic structure in PDB format as input, can be used with experimentally determined structures as well as structural models (predicted structures, Sect. 2.5).

2.3.1 Structure Manipulation and Visualization

Almost all bioinformatics studies including protein 3D structures involve, at some point, manipulating and visualizing them.

The best software for visualizing and manipulating protein structures comprises stand-alone programs which, while not difficult to use, do not run within a web browser and have to be locally installed in an operative system dependent manner. Consequently, they do not fall within the scope of this book. Nevertheless, due to the large difference in capabilities with the browser-based solutions we are going to comment here, we recommend readers particularly interested in this subject to explore these solutions, for example PyMol (http://www.pymol.org/).

The most widely used molecular visualizer designed to run within a web browser is ***JMol***. JMol runs as a Java applet embedded in a web page together with other elements. Normally, JMol applets show up "preloaded" with a 3D structure, for example in the web page of the corresponding entry for that structure in the resources commented in Sect. 2.2. But it can also be used to load files with structures provided by the user (e.g. a predicted structure). Additionally, JMol is highly customizable (size, contents of the menus, etc.) and can be connected with other elements in the web page, which makes different JMol applets to look slightly different, depending on the web page they are embedded and the purpose of the molecular visualization in that particular page. Although, the web address for JMol provided below is that

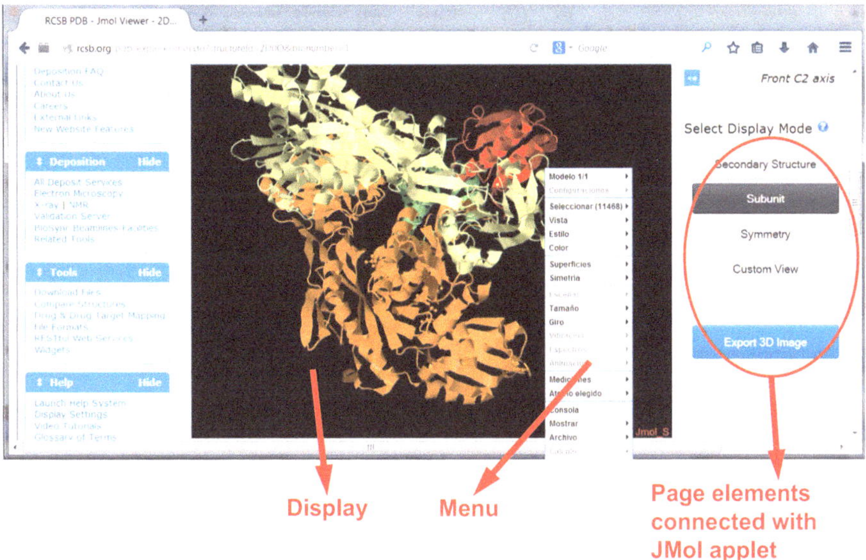

Fig. 2.4 JMol applet embedded in a PDB entry page at the RCSB

of its main page, which contains the description of the system, documentation, etc., the JMol applet embedded there (configured mainly for demo purposes) is not the most useful one for general purposes. Consequently, in the following we are going to describe a particular example of JMol, that implemented in the PDB resource described in Sect. 2.2.

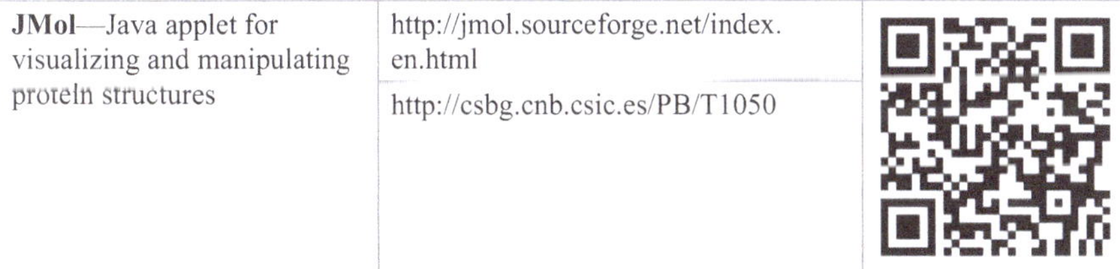

JMol—Java applet for visualizing and manipulating protein structures	http://jmol.sourceforge.net/index.en.html	
	http://csbg.cnb.csic.es/PB/T1050	

Note: In recent versions of Java, in order for the embedded applets to run, you have to set the security level to "middle" in the Java configuration panel of your operative system (e.g. in Windows: Control panel > Java > Security > security level > middle). Additionally, the first time the applet is run, you will have to accept a number of security warnings and "Allow…?" questions.

As commented in Sect. 2.2, the "3D View" tab of the PDB entry for a given structure includes a JMol applet to interactively visualize and manipulate it (Fig. 2.4). By default, the protein chains within that entry are shown in "ribbon" representation so as to highlight the secondary structure elements, which are colored pink (α-helices), yellow (β-strands) and purple (3.10 helices). The ligands are shown in "ball & stick" representation. The structure can be rotated dragging the mouse while the left button is pressed. Dragging with the middle button pressed (or using the mouse wheel) zooms in/out. Dragging with the CTRL key and right button pressed moves (translates) the representation.

The right button alone pops up JMol's menu. The type of representation (ribbon, surface, spheres, wireframe …) can be selected in "Style>Schema". The "Spin" menu can be used to automatically rotate the molecule without user intervention. The submenus of this item control the axis to rotate around, the speed, etc. When the mouse stays over an atom for a while, a yellow box with its label (residue number, atom type …) appears. For labeling residues in a permanent way as you click on them, use "Set picking>Label". For stop labeling on click use "Set picking>Off".

An important concept in JMol is that of "selection". A selection is a subset of the atoms within the structure so that all forthcoming operations will be applied to them only. This allows representing different parts of the structure in different colors, styles, etc. so has to highlight them. The "Select" menu contains a number of pre-defined subsets such as atoms in proteins, in nucleic acids, in ligands, etc. For example, to highlight the ligands of the structure (except waters) in green color you do: "Select>Hetero>Non-aqueous HETAM", then "Style>Scheme>CPK", and finally "Color>Atoms>Green". For arbitrary selection of residues you can use "Set picking>Select group". This will add to the current selection the residues you click. "Set picking>Off" to stop selecting on click. When performing complex selections, it is recommended to temporary mark them with "Select>Selection halos", so as to visualize in real time what you have in your selection. With "Select>Display selected only" it is possible to remove from the representation all atoms not included in the current selection. Use "Select>All" to go back to the initial state in which the operations affect all atoms in the structure.

The "Measurements" menu can be used to measure distances and angles. For example, to show the distance between two atoms, use "Measurements>Click for distance measurements". This will start showing the distance for every pair of atoms clicked since then. To stop: "Set picking>Off". The "File" menu allows opening PDB files locally stored in the user's computer. It also has a number of options to export the current representation as an image file or as a 3D scene (to be imported in other 3D software). Another interesting option of the "File" menu is "Save script with state", which allows saving a session file that can be used to restore the current representation later, in forthcoming sessions of JMol (with "File>Open script).

Most pages with JMol embedded have a number of pre-build scripts which can be sent to JMol just clicking a link or button, allowing in this way to generate complex pre-defined representations. For example, in the "3D View" tab of a PDB entry we are discussing, there are a number of buttons on the right to change the global representation (color by chain, by secondary structure, etc.) (Fig. 2.4). There are also two tabs at the bottom of the page ("Ligands" and "Domains") which contain links to highlight in the JMol applet the different domains of the chains, ligand binding pockets and residues involved in binding, etc.

2.3.2 Structure Comparison

The methodologies for structurally aligning protein structures allow performing the same basic operations we have seen for sequences (Chapter 1). That is, align and compare two protein structures (pairwise alignment), align a set of structures (multiple alignment) and finding similar structures in a database (database search).

Pair-Wise Structural Alignment (Comparing Two Structures)

Aligning two structures is useful, for example, to compare two instances of the same protein crystallized in different conditions or with different cofactors so as to understand the structural changes associated to these differences. It is also useful for detecting the structural differences between two homologous proteins in order to relate those to the functional differences and infer the protein regions associated to them. Structural alignment is also the only way to find equivalent residues between two homologous proteins when their sequences are too divergent for sequence-based alignment methods. Indeed, structural alignments are used as "gold standards" to benchmark sequence alignment methods and derive amino acid exchange matrices for sequence comparison (Sect. 1.5). Since the methodologies for structural alignment are not as established as those for sequence alignment, there are many different approaches available, with their associated web servers. Nevertheless, for cases of clear structural similarity (even if local) the results are quite similar.

Apart from the explicit structural superposition, these methods also provide various figures which try to quantify the goodness of the structural alignment. Usually, these include the "root mean square deviation" (RMSD), which quantifies the average distance between equivalent atoms and the percentages of structurally equivalent residues. As with sequence comparisons (Sect. 1.6.2), a statistical estimator of the likelihood of the structural match (p-value or z-score) is also provided in most cases.

If you want to compare two protein structures which are deposited in PDB and for which you know the PDB IDs, you can use the ***RCSB-PDB protein comparison tool*** (Prlic et al. 2010).

RCSB-PDB protein comparison tool—Structural alignment of two proteins (or domains) deposited in PDB	http://www.rcsb.org/pdb/workbench/workbench.do http://csbg.cnb.csic.es/PB/T1060	

The inputs for this tool are the IDs of the two structures you want to compare, and the method you want to use for the structural alignment. There is an "auto-suggest" feature so that a list of possibilities show up when you start typing your ID. For

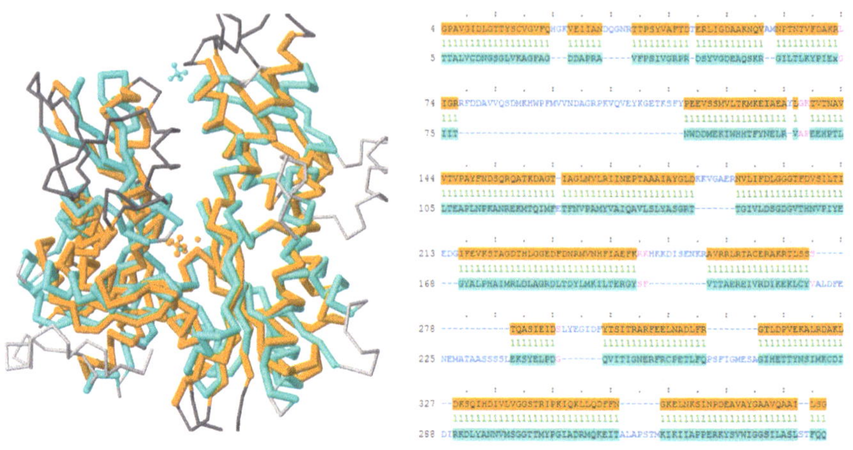

Fig. 2.5 JMol applet showing the structural alignment between hexokinase (PDBID_chain: 3hsc_A *–orange-*) and actin (3hbt_A *–blue-*), two distant homologs

example, if you type a PDB ID of a structure with multiple chains, these appear as the possibilities for you to choose. The SCOP structural domains associated to that PDB ID also appear, which opens the interesting possibility of aligning individual domains, instead of whole chains. The "Select comparison method" menu contains not only structure alignment but also sequence alignment tools (Sect. 1.5). Regarding the structure alignment methods, maybe the two most interesting ones are "jCE" (Shindyalov and Bourne 1998) for rigid alignment and "jFATCAT-flexible" (Ye and Godzik 2003) for flexible alignment. In flexible structural alignment, some movements of the protein chains are allowed to take into account cases of similar structures but "distorted" by whatever reason (crystallization conditions, domain movements …). Once the two protein chains and the alignment method are selected, you can press "Compare".

Results vary slightly depending on the selected comparison method, but in general they include a JMol applet (Sect. 2.3.1) to visualize/manipulate the two structures aligned, highlighting the aligned segments of the two proteins with different colors (Fig. 2.5). The results also include the implicit sequence alignment (to highlight the equivalent residues) and some scores associated to the structural alignment, such as the RMSD and the percentage of sequence identity according with the implicit alignment. There is also a link to download a PDB file with the two structures aligned, for further manipulation. In this PDB file, the two aligned chains are stored as two models ("MODEL 1" and "MODEL 2") in the format used for representing NMR-derived structures comprising multiple models.

The ***Dali_lite*** server at the University of Helsinki allows to structurally align not only two protein chains deposited in PDB for which you know the PDB IDs, but any pair of structures you have locally stored in PDB files as well (Holm et al. 2006).

Dali_lite—Structural alignment of two protein structures	http://ekhidna.biocenter.helsinki.fi/dali_lite/start	
	http://csbg.cnb.csic.es/PB/T1070	

In the input form, you have to enter the PDB IDs of the two proteins to align, or upload the two PDB files in the "mol1" and "mol2" sections. If the PDB entries or local files comprise multiple chains, you have to specify which chains to align in the corresponding boxes. After pressing "Submit", in a few seconds you are taken to a page with the results of the structural alignment. The most important information of the "Summary" section is the RMSD of the structural alignment and the percentage of sequence identity it renders. In the "Pairwise structural alignment" section (bottom of the page) the by-residue alignment is shown, highlighting the identities and showing the secondary structures of both chains. Lowercase residues are those without structural equivalent in the other chain. To visualize the structural alignment in JMol, mark the checkbox within the "Summary" section and press the "3D superimposition—JMol applet" button above. Although Dali_lite does not provide a downloadable PDB file with the two structures aligned, it is possible to download a PDB file with the second structure ("mol2") rotated/translated according with the alignment ("PDB" link of the "Summary" section). So, to visualize the structural alignment outside Dali_lite, you simply have to load the original PDB file of the first structure, and then this rotated/translated PDB of the second.

Looking for Structural Homologues (Finding Similar Structures in a Database)

Detecting which proteins are structurally similar to one of interest can provide functional and evolutionary information about the latest. Even if structurally similar proteins can, in general, have different functions and evolutionary origins (Sect. 2.2.1), some particular folds are functionally homogeneous. In these cases, a structural match of our protein of interest against them can be directly interpreted in functional terms. Realizing that our protein has no structural homologs in databases ("novel fold") is also interesting from many points of view.

Protein structure databases usually have pre-compiled lists of structural homologs for their entries. These are updated as new entries are added. Structural comparison is a CPU-intensive problem. Additionally, close homologs (proteins with high sequence identity) have virtually the same 3D structure and, consequently, it does not make sense to perform a structural alignment with all of them. For these reasons, these pre-compiled structural comparisons usually do not include all entries (protein structures) but only a representative structure for each "sequence cluster" (group of close homologs), and tools are provided to transfer the structural alignments to any of the other members of the group (Fig. 2.6).

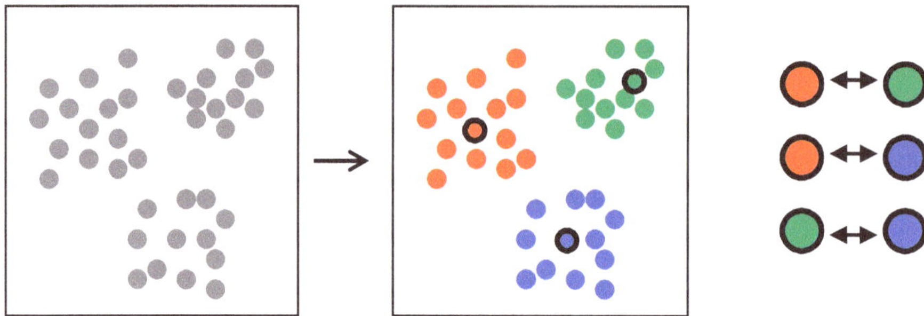

Fig. 2.6 Generation of precompiled lists of structural homologs. The grey dots represent the proteins of known structure deposited in a given database, and they are arranged in a sort of "sequence space" where the distances between them are (*inversely*) proportional to their sequence identities. Clusters of similar sequences are defined according with a given threshold of sequence identity (*colors*). A representative protein is selected for each cluster (*wider circles*) and the structural alignments (*right panel*) are performed for them only

This is the case of the ***RCSB***, where the ***"3D similarity" tab*** within the page of a given PDB entry provides information on its structurally similar proteins. In this case, the sequence clusters (Fig. 2.6) are defined based on a sequence identity cutoff of 40 %. Unless the PDB entry we are interested in is the representative of its cluster, the "3D similarity" tab shows this representative structure of the 40 % cluster where our entry is, as well as the sequence identity between both. A link is provided to retrieve the precompiled results of the structural search of that representative against those of the other clusters.

The page with the results of a structural search is quite similar in format to a typical sequence search (e.g. BLAST, Sect. 1.6.2): a list of homologs (structural homologs in this case) sorted by some score (p-value). For each of these structural matches, some parameters of the corresponding pair-wise structural alignment are provided, such as the RMSD, the coverage in both domains, the percentage of sequence identity associated to the structural alignment, etc. Information on these parameters is provided in the "Table Legend" column on the right. There is also a link ("view") to visualize these pair-wise structural alignments and obtain additional data on them.

The ***Dali database*** at the University of Helsinki (Holm and Rosenström 2010) can also be used for retrieving the (pre-calculated) structural homologs of a given protein deposited in the PDB. As with the RCSB similarity tool described above, protein chains within PDB are clustered based on sequence identity (90 % in this case) and structural similarities are calculated only for representative structures of each cluster.

Dali database—Precompiled structural homologs for a protein deposited in PDB	http://ekhidna.biocenter.helsinki.fi/dali/start http://csbg.cnb.csic.es/PB/T1080	

In the input form, only the PDB ID and the chain identifier of the protein for which you want to retrieve the structural homologs are required. The format of the results is identical to those of the Dali_lite pair-wise structural alignments commented above, with the difference that now there is a list with (eventually) many structural homologs, instead of just one. Consequently, many of the operations described above for Dali_lite can now be performed for many proteins (e.g. select many structural homologs –checkboxes-) to show the (multiple) structural alignment in JMol.

The ***Dali server*** (Holm and Rosenström 2010) is similar to the Dali database but it allows using any structure as input, either deposited in PDB or not (e.g. a predicted structure).

Dali server—Retrieve structural homologs for a generic 3D structure	http://ekhidna.biocenter.helsinki.fi/dali_server http://csbg.cnb.csic.es/PB/T1090	

In this case the input form contains a button to upload the local file with the structure (in PDB format). Since in this case structural similarities are not pre-calculated and, consequently, the running time can be quite long, the user can enter an e-mail address to be notified upon job completion. The result pages are identical to those of the Dali database.

The ***PDB eFold*** tool at the EBI (Krissinel and Henrick 2004) also allows, among many other things, to retrieve the structural homologs of a given protein. The main difference with the resources described previously, and the reason for including it here, is that the search is not restricted to a representative set of structures, but it can be performed against the whole PDB, in a similar way BLAST sequence searches can be done against whole sequence databases and not only non-redundant versions of them. This can be useful if, for example, we suspect that the assumptions commented earlier about the structural similarity between a representative structure and other proteins in its cluster do not hold, or if we simply do not want to transfer structural features and/or alignments between them and prefer a direct comparison.

PDB eFold—Structural searches against the whole PDB. Also pairwise structural alignment and multiple structural aligment.	http://www.ebi.ac.uk/msd-srv/ssm/ http://csbg.cnb.csic.es/PB/T1100	

In the submission form (pairwise mode) you select your source structure (PDB/SCOP entry or local file with 3D structure) and the set of structures to compare against. If this set contains a single structure (PDB/SCOP entry or local file) the server will perform a pair-wise structural alignment. For searching against multiple structures, the set can contain the whole set of PDB chains or SCOP domains. For the input structure, we can restrict the structural search to a given chain or individual domain, or even a range of residues within it ("select chains" pop-up menu). The rest of parameters in the form are ok for most situations.

Once the query was processed, the results page is quite similar to those described above: a list of structural matches for our input structure and their associated parameters (RMSD, % sequence identity …) as well as links to the entries of these matches in different databases. The links of the first column ("##") allow retrieving additional data on that particular structural match, including detailed residue-by-residue information on the alignment, as well as the possibility of inspecting it in JMol and other viewers.

Multiple Structural Alignments (Comparing More than Two Structures)

In some circumstances, it could be useful to structurally align a large set of proteins. For example to highlight functionally important structural differences between groups or to look for conserved positions in a set of remote homologs too divergent to be aligned by sequence-based methods.

Most servers for the search of structural homologs discussed above offer the possibility of generating a pseudo-multiple structural alignment with the found structures. It is "pseudo-multiple" because it is just a piling-up of all the binary pair-wise alignments of the found structures against the query. If we need a "really-multiple" structural alignment, or want to align our own set of structures not coming from a structural search, there are some servers which offer that possibility.

The ***PDB eFold*** tool described above can generate multiple structural alignments (Krissinel and Henrick 2005) with the "multiple" option at the top of its web form. The individual PDB files with the structures to be aligned should be packed in a single ".tar" file. In MS-Windows, these files can be created with free software such as 7-Zip (www.7-zip.org), and in UNIX-based systems, such as Linux or Mac OSX, with the "tar" command. Once this file is created, we select "File set" in the "Source" menu and upload it. Pressing "Update List", the "List of entries" box on the left of the form becomes filled with the different structures packed in the ".tar" file. If this list is correct, we can press "Submit query" to start the alignment.

The results of the multiple structural alignment are similar in format to those for pair-wise alignments, but extended to multiple structures. The "Overall RMSD" parameter reflects the overall structural similarity of the whole set of structures. At the bottom of the page there is a by-residue representation of the multiple alignment. The sequence alignment determined by the structural alignment can be downloaded (multiple FASTA format) with the "download FASTA alignment" button. Finally the multiple structural alignment can be visualized in 3D in a JMol applet with the "view superposed" button.

2.4 Prediction of 1D Structural Features

In some cases a prediction of the whole 3D structure of a protein (Sect. 2.5) is not possible or not necessary, and predicting "low-level" structural characteristics is enough and useful.

One dimensional (1D) features are those structural characteristics of a protein that can be represented as individual values (either qualitative or quantitative) associated to the residues of the protein (individually, but most commonly in consecutive sequence segments). For example: secondary structure, exposed/buried residues or transmembrane segments. Mapping the predicted 1D features along the sequence of our protein is very useful for having a first ("low resolution") map of its overall structural characteristics and topology.

2.4.1 Secondary Structure and Solvent Accessibility

Although there are many more secondary structure (SS) elements, most programs are designed to work with a 3-state alphabet: alpha-helix (H), beta-strand (E) and the rest (C or '-', depending on the program).

The first generation methods were based on single amino acid propensities to form the three types of SS elements, extracted from known structures. Current methods combine information on single amino acid, segments of consecutive residues, possible nonlocal relationships (like the molecular interactions established in beta-sheets) and conservation (i.e. considering not a single sequence but a multiple sequence alignment) so as to achieve much higher levels of accuracy. Since the inner working of modern methods for predicting secondary structure and solvent accessibility is very similar (based on training with known examples) and so is the input they use, almost all tools predict these two types of 1D features concomitantly.

In general the input for these methods is a MSA of homologous sequences (inferred to have the same 3D, and hence secondary, structure) but almost all methods can take a single sequence as input and internally generate a MSA for it automatically. Nevertheless, if you have a manually-curated MSA for your protein of interest, it is better to use it for predicting these 1D features instead of relying on that automatically generated.

A widely used server for predicting secondary structure and accessibility is ***JPred***, (Cole et al. 2008) hosted at the University of Dundee.

JPred—Prediction of secondary structure and solvent accessibility	http://www.compbio.dundee.ac.uk/jpred/ http://csbg.cnb.csic.es/PB/T1110	

In the web form of JPred, you can paste your protein sequence and click on "Make Prediction".

If a similar sequence to that you want to analyze is found in the PDB, JPred provides a link to the corresponding PDB entry, so you have access to the "real" secondary structure and accessibility, instead of predictions.

You can also follow the "Advanced" link in the main form to control some input parameters, such as providing your own MSA file and skip searching PDB for homologous proteins of known structure. You can also provide an e-mail address to receive the results and a name for the job.

Upon successful completion of a job, JPred delivers the results in various formats: there is a button to launch the alignment viewer Jalview preloaded with the alignment internally generated for your sequence and the predicted 1D features as "annotation" lines (Sect. 1.7.2). The predictions are also available as printable documents (PS and PDF). Additionally, you have access to the intermediate files generated during the prediction ("raw data") some of which can be of interest: for example the file containing the multiple sequence alignment internally generated in FASTA format.

In the following, we are going to comment the "HTML full" format for the results. This file starts with the alignment generated for your query sequence. The IDs of the proteins are active links to the corresponding database entries. At the bottom, you can find the secondary structure predicted by three different methods ("Jnet", "jhmm" and "jpssm" lines). The predictions where the three methods agree are obviously more reliable.

JPred also provides prediction of solvent accessibility as buried (B) or exposed (-) residues (see "Jnet_25", "Jnet _5" and "Jnet_0" lines for different accessibility thresholds. It also provides coiled-coil prediction in the "Lupas lines" (Sect. 2.4.3).

JPred is also conveniently included in the MSA editor program Jalview (Sect. 1.7.2). To predict the secondary structure of a protein from within Jalview, select its sequence in the MSA and, from the alignment window menu, choose "Web service → Secondary structure prediction → JNet secondary structure prediction". A new window will show the MSA automatically constructed together with various annotation lines at the bottom: "Lupas" for coiled-coil predictions, "JNETSOL" for solvent accessibility and "JNET" for secondary structure predictions.

Another widely used predictor of secondary structure and solvent accessibility is ***PSIPRED*** (Jones 1999), at the University College London.

PSIPRED—Prediction of secondary structure and solvent accessibility	http://bioinf.cs.ucl.ac.uk/psipred/ http://csbg.cnb.csic.es/PB/T1120	

In the input form, make sure that "PSIPRED (predict secondary structure") is selected as "prediction method". Then paste your sequence or a multiple sequence alignment (FASTA format). As with JPred, if a single sequence is provided the server automatically generates a MSA for it. Although not required, an email address for being informed of job completion is recommended since this server can take a long time to run. A job identifier is required.

After job completion, you are taken to a results page with three tabs. The "summary" tab contains a linear representation of your sequence highlighting the regions predicted as forming alpha-helix (pink) and those predicted as beta (yellow). The "PSIPRED" tab contains a printable version of the prediction, including a representation of the by-residue reliability (blue bars). Finally, the "Downloads" tab contains links to the predictions in additional formats, such as plain text and PDF.

2.4.2 Transmembrane Segments

The structures of integral membrane proteins that have been already solved are classified in only two groups, depending on the secondary structure elements used to span the membrane: alpha-helical bundles or beta-barrels (Neumann et al. 2010). Luckily the environment defined by the chemical composition of biological membranes imposes strong restrictions in the amino acid composition of the transmembrane (TM) segments of protein sequences. For this reason, the prediction of regions in protein sequences that might be traversing biological membranes is rather accurate. Obtaining a prediction of transmembrane segments for a protein sequence allows generating a topological map (von Heijne 2006) in which the intra- and extra-cellular domains become evident. Such topological map, apart from the intrinsic utility of providing a first overview of the organization of the protein in structural/functional domains (e.g. Fig. 2.7), can be used for example to "trim" these domains so as to express and crystallize them without the TM parts.

There are two types of programs to predict transmembrane segments: programs in the first group assume alpha-helical bundle conformation, while programs in the second assume proteins to adopt a beta-barrel conformation. A prior prediction of secondary structure (previous section) may help in deciding which type of TM predictor to use.

TMHMM predicts transmembrane helices in proteins together with membrane topology from a single protein sequence (Krogh et al. 2001).

TMHMM—Prediction of transmembrane helices and transmembrane topology	http://www.cbs.dtu.dk/services/TMHMM-2.0 http://csbg.cnb.csic.es/PB/T1130	

To use the server, upload a sequence file or paste it in the form (FASTA format), and click the "Submit" button.

As an example, Fig. 2.7 shows the TMHMM prediction for the following sequence:

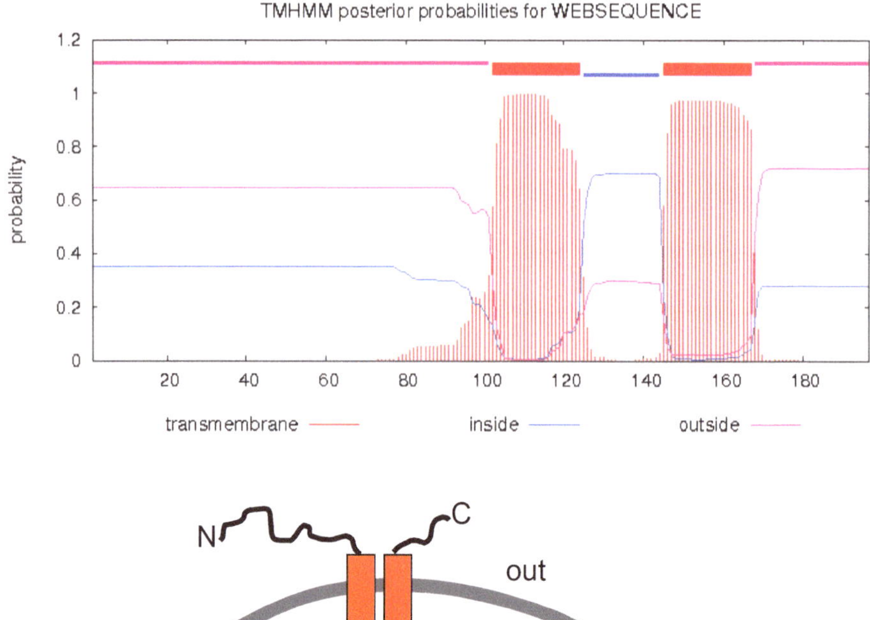

Fig. 2.7 TMHMM prediction for E9CFG9 and the topological organization inferred from it

```
>UniRef90_E9CFG9 ER membrane DUF1077 domain-containing protein n=1
Tax=Capsaspora owczarzaki (strain ATCC 30864) RepID=E9CFG9_CAPO3
MAHKRDARFTLDLARTLPQDSTATDASGSSSSSTSISTSNNSGSGELLSGYDATFRGSAD
ARTARNPEALAQLKSMKAWEMALAPAKSVPMNGFMMWMAGNSIHIFSIMITVMMLITPVK
AIFSTGTTFAKLTEDGKSQLLQQKLVFILANCLSIGMAMYKFSVLGLLPTSPSDWLSFLD
PKQILEVSVGSTAAAPM
```

Complemented with the following textual output:

```
# WEBSEQUENCE Length: 197
# WEBSEQUENCE Number of predicted TMHs:  2
# WEBSEQUENCE Exp number of AAs in TMHs: 45.74295
# WEBSEQUENCE Exp number, first 60 AAs:  0.00028
# WEBSEQUENCE Total prob of N-in:        0.35300
WEBSEQUENCE TMHMM2.0    outside          1    101
WEBSEQUENCE TMHMM2.0    TMhelix        102    124
WEBSEQUENCE TMHMM2.0    inside         125    144
WEBSEQUENCE TMHMM2.0    TMhelix        145    167
WEBSEQUENCE TMHMM2.0    outside        168    197
```

TMHMM predicts two transmembrane helices: one in the 102–124 region and the second in the 145–167 region. It also predicts the topology of the protein sequence across the membrane, defining the outside and inside domains from the prediction of the localization of its N-terminal (outside in this case). With this single prediction, it is possible to generate a first topological draft of the protein (Fig. 2.7). The structural characterization of this protein could go on, for example, with the prediction of structure for the N-terminal (extracellular) long domain.

Phobius (Käll et al. 2007) can predict simultaneously signal peptides and transmembrane helices from a single protein sequence ("Normal prediction"), or from a multiple sequence alignment ("PolyPhobious").

Phobius—Simultaneous prediction of signal peptide and transmembrane helices	http://phobius.sbc.su.se/ http://csbg.cnb.csic.es/PB/T1140	

For predicting from a single sequence, paste it (in FASTA format) and press "Send". For predicting from a MSA, follow the "PolyPhobious" link in the Phobius web site. Paste your MSA or upload the MSA file (in FASTA format) and select "Aligned fasta" as input format. Finally click the "Send" button.

In addition to transmembrane helices, Phobius reports also predicted signal peptides (Sect. 2.4.5)

BOCTOPUS (Hayat and Elofsson 2012) predicts potential transmembrane segments with a beta barrel topology.

BOCTOPUS—Prediction of transmembrane strands	http://boctopus.cbr.su.se/ http://csbg.cnb.csic.es/PB/T1150	

For using the server, paste your query sequence or upload a file with it (FASTA format) and click the "Submit" button.

Figure 2.8 shows BOCTOPUS predictions for protein C4U1J5. As you can see, the server predicts the N-terminal of this protein to be in the cytosol followed by a transmembrane beta barrel with 12 strands, so that the C-term ends in the cytosol again.

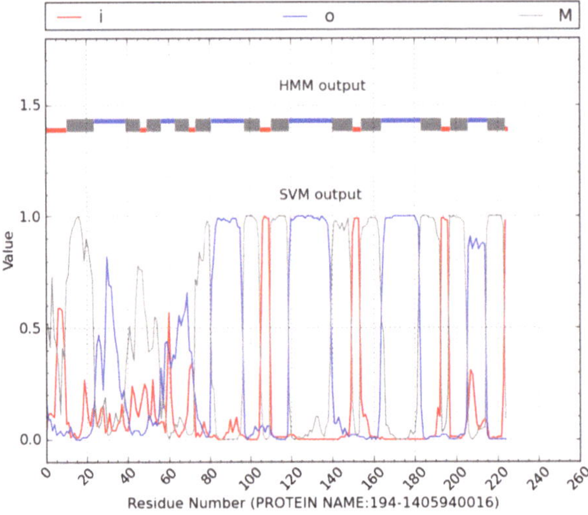


```
>194-1405940016
iiiiiiiiiiMMMMMMMMMMMMMMooooooooooooooooooMMMMMMMiiiMMMMMMMoooooooMMMMMMM
iiiMMMMMMMMooooooooooooooooMMMMMMMMiiiiiMMMMMMMMMooooooooooooooooooooo
MMMMMMMMMMiiiiMMMMMMMMMMooooooooooooooooooooooMMMMMMMMMMiiiiMMMMMMMMMoooo
oooooMMMMMMMMMi
```

Fig. 2.8 BOCTOPUS predictions for C4U1J5

2.4.3 Coiled-Coils

Coiled-coils are a type of alpha-helices with a particular periodic amino-acid composition which makes them wound into superhelical structures. These super-helices can be formed by different numbers of alpha-helices twisted around each other. These alpha-helices can come from the same polypeptide or from different ones (inter-protein coiled-coil) and, consequently, these structures are frequently mediating protein-protein interactions. For this reason, the main utility of predicting coiled-coils in our sequence, besides adding that information to the general topological map together with the other 1D-predictions, is to infer protein interaction segments (Fig. 2.9).

As with TM segments, the particular and periodic amino-acid composition of these helices makes them easily apprehensible for prediction methods, consequently rendering good accuracies.

COILS (Lupas et al. 1991) predicts the presence and exact localization of coiled-coils in a protein sequence.

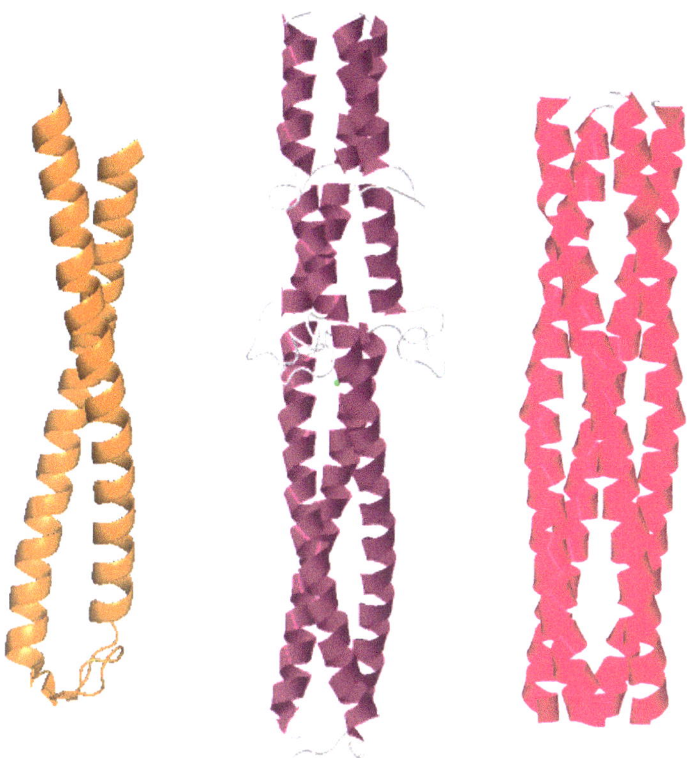

Fig. 2.9 Examples of three types of coiled-colis: dimeric (PDB 1l8d – *orange*-), trimeric (PDB 1aa0 –*purple*-) and tetrameric (PDB 1nhl –*pink*-)

COILS—Prediction of coiled-coils	http://www.ch.embnet.org/software/COILS_form.html	
	http://csbg.cnb.csic.es/PB/T1160	

In the COILS input form, paste your protein sequence (in plain text, with no header line) and press "Run coils". COILS generates a plot with the probability of finding a coiled-coiled along the length of the sequence, using three window sizes for scanning the sequence (14 in green, 21 in blue and 28 in red). In general, the predictions based on the three window sizes should agree for long coiled coils. The short window predictor (14) can be used as an indicator of short coiled-coils (providing the score is good) that by definition will never render good scores in the other two windows sizes.

The server also generates a text file ("numerical format") containing the scores for each individual residue together with an assignment to each of the (***a*,** b, c, ***d***, e, f, g) positions in a canonical hepta-repeat. Coiled-coils are characterized by the repetition of that 7-residue motif where the 1st and 4th residues (***a*** and ***d***) are hydrophobic and form a hydrophobic "velcro strip" along the helix responsible for the binding of the other helix. Consequently, knowing which residues are in each position of the motifs allows, for example to locate these responsible for the binding, design mutants, etc.

The ***LOGICOIL*** server (Vicent et al. 2013) at the University of Bristol predicts not only coiled-coil regions in our query sequence, but also their potential oligomeric state as well.

LOGICOIL—Prediction of coiled-coil regions and their oligomeric state	http://coiledcoils.chm.bris.ac.uk/LOGICOIL/	
	http://csbg.cnb.csic.es/PB/T1170	

Select input format "OPTION II: whole protein sequence" and increase the "Marcoil threshold value" to 50, for example. Finally paste your sequence (in FASTA format) and press "Submit".

In the results plot, make sure that the threshold you have introduced (50) defines the coiled coil segments you intuitively see in the plot and, eventually, go back in your browser and run the server again with another threshold. The number of coil segments defined by this threshold would affect the forthcoming oligomerization predictions. For example, in Fig. 2.10 we see a clear coil segment from residues 20 to 80. A threshold of 50 (horizontal line in Fig. 2.10) would also define this 1-coil prediction and, consequently the system will try to predict the multimerization state for it ("trimer" in this case). Nevertheless, a threshold of 90 would define two coil segments, what would affect the oligomerization prediction. If we have evidences that this is the case and there are two coils instead of one (for example coming from other predictors) we can move the threshold accordingly. Indeed, the "OPTION I" of the input form allows to import a generic coiled-coil prediction (with "abcdefg" assignments) and the oligomerization prediction will be based on it.

2.4.4 *Disordered Regions*

Disorder is a general term that refers to intrinsically flexible regions within proteins, which might extend in some cases to the full length of the protein sequence. These regions are totally or partially unfolded in their native form, and in spite of previous beliefs, are implicated in various and important functions including, but not limited to, molecular interactions and regulation.

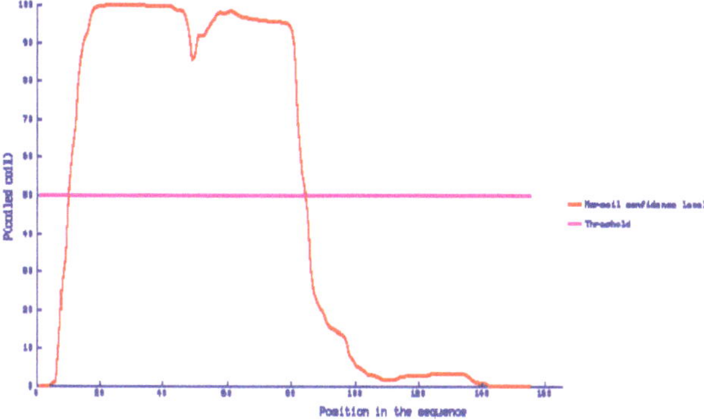


```
MARCOIL predicted region: 1
Sequence:DYAKRLGDLEGEVKDLRQKVDDLARGYQDILKELLQLQKGLEAKADKKETERELESLSRLV
MAVYDKIKELETVN
Register:cdefgabcdefgabcdefgabcdefgabcdefgabcdefgabdefgabcdefgabcdefga
bcdefgabcdefga
Result of prediction: Most probable state is TRIMER
Second most probable state TETRAMER
            ANTI PARA TRIM TETRA
Raw score is 0.93 1.05 1.34 1.1
```

Fig. 2.10 LOGICOIL prediction for L0B8L1. The threshold has been set to 50 so that it defines 1 coil segment. The output includes an assignment of residues to the (abcdefg) repeats, and the residues predicted to form the hydrophobic strip (a and d) are highlighted

The utility of detecting disordered regions in our sequence is twofold. In one hand, it contributes, together with other 1D features, to delimit the topological map of our protein and point to functional regions (interaction, flexible linkers, "springs" …). In the other hand, it allows to design constructs where these segments are excluded so as to express and eventually crystallize the globular domains. Many attempts to determine the structure of proteins failed due to the presence of (undetected) long disordered segments.

There is a certain relationship between disordered segments and sequence regions of "composition-bias", the type of regions masked by some BLAST filters (Sect. 1.6.2). Consequently, the presence of masked segments in a blast search (labeled with "X") can be a first indicator of disordered segments, which should be further confirmed with the specific predictors commented below.

A widely used disorder predictor is ***DISOPRED***, at the University College London (Ward et al. 2004).

DISOPRED—prediction of disordered regions	http://bioinf.cs.ucl.ac.uk/psipred/?disopred=1	
	http://csbg.cnb.csic.es/PB/T1180	

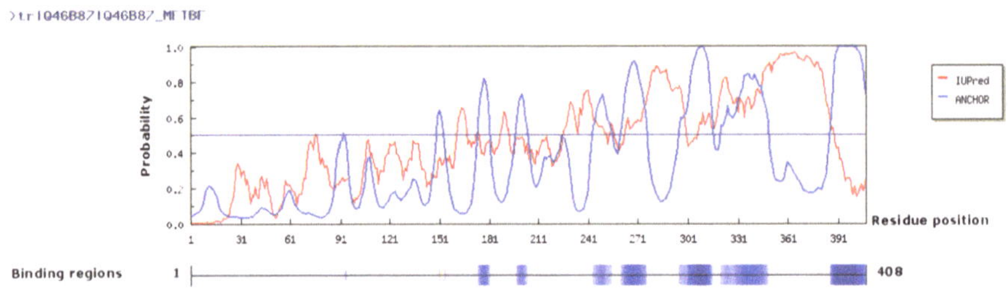

Fig. 2.11 IUPred (*red*) and ANCHOR (*blue*) predictions for Q46B87

In the web form, check that "DISOPRED3 & DISOPRED 2 (Disorder Prediction)" is selected in "Choose Prediction Methods", paste a single sequence or a multiple sequence alignment (both in FASTA format), provide a short label for your request in "Short identifier for submission" at the bottom of the page, and finally click on "Predict". It might take a long time for DISOPRED to run. For this reason, it is recommended to enter an email address to be informed of job completion.

The result pages are almost identical to those of PSIPRED (Sect. 2.4.1). In this case, residues predicted to be in disorder state are marked with a red box and those predicted to be disordered and involved in binding with a green box.

Another good predictor of disorder is ***IUPred*** (Dosztanyi et al. 2005), at the Hungarian Institute of Enzymology.

IUPred—Prediction of disordered regions	http://iupred.enzim.hu/ http://csbg.cnb.csic.es/PB/T1190	

In the form, enter you protein sequence (in raw or FASTA format) or its UniProt accession number, select "Prediction type" (generally you can use the default "long disorder") and click the "SUBMIT" button. By default, IUPred shows results in a graphical plot together with a table of prediction values per residue. In the plot, the threshold order/disorder is indicated with a horizontal line. Consequently regions above that line are predicted as disordered.

Figure 2.11 shows the IUPred disorder predictions for the protein with Uniprot accession Q46B87 (red line). It can be seen that the C-terminal half of this sequence (from residue 230) is predicted as disordered.

ANCHOR (Dosztanyi et al. 2009), from the same developers of IUPred, is designed to predict disordered regions potentially involved in protein-protein interactions.

ANCHOR—Prediction of disordered regions involved in protein interactions	http://anchor.enzim.hu/ http://csbg.cnb.csic.es/PB/T1200	

The input form and the results page are quite similar to those of IUPred. Indeed, the IUpred prediction is included in the results plot (red line in Fig. 2.11) together with ANCHOR's prediction (blue line in Fig. 2.11). As for IUPred, regions with an ANCHOR score above the threshold are predicted as disordered regions involved in protein interactions.

In the example of Fig. 2.11, it can be seen that within the C-terminal half of the protein predicted to be disordered, there are a number of segments (blue boxes) predicted to be involved in interacting with other proteins.

2.4.5 Protein Sorting Signals

SignalP (Petersen et al. 2011), at the Center for Biological Sequence Analysis of Denmark, predicts whether there is a signal peptide and its exact location (its cleavage site) in protein sequences from different taxonomic groups, namely Gram-positive bacteria, Gram-negative bacteria and eukaryotes.

SignalP—Prediction of signal peptides	http://www.cbs.dtu.dk/services/SignalP/ http://csbg.cnb.csic.es/PB/T1210	

In the Submission section of SignalP paste the protein sequence in FASTA format. Select the appropriate group for the organism (Eukaryotes, Gram-negative bacteria or Gram-positive-bacteria), and press "Submit". If you know that your sequence does not contain transmembrane helices, you can select "Input sequences do not include TM regions" in the Method section to increase the predictor performance.

Figure 2.12 shows the SignalP predictions the protein with Uniprot accession G3VVG6. SignalP shows in a graph three scores (C, S and Y) along the sequence length. The C-score (red) is high for the residue(s) predicted to be immediately after the cleavage site (the first residue in the mature protein). The S-score (green) is high along the predicted signal peptide. And the Y-score (blue) is a combination of the C- and S-scores.

The output also includes the following text with the position and maximum value of each score, as well as the mean S-score. Finally SignalP reports D, a global summary score that indicates whether a signal peptide is present.

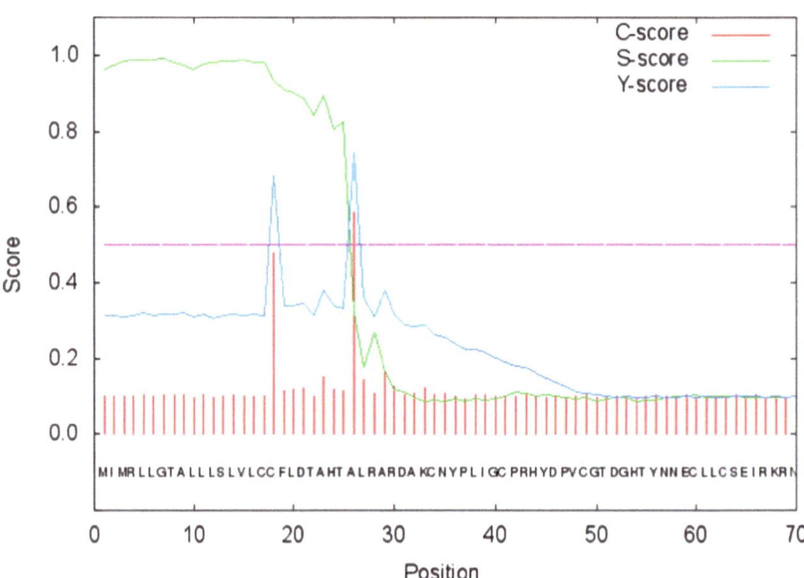

Fig. 2.12 SignalP predictions for G3VVG6

```
SP='YES' Cleavage site between pos. 29 and 30: VTT-EI D=0.713 D-cutoff=0.450
Networks=SignalP-noTM
```

Taking together, these predictions indicate the presence of a signal peptide in the first 29–30 residues of that protein. A table with the C, S and Y scores can be obtained from the "data" link.

2.5 Predicting Protein 3D Structure

A 3D structural model of a protein can be tremendously useful. If it is reliable enough, it is almost like having the experimental structure, and you can apply almost all analysis methods described in Sect. 2.6 for obtaining information from that structure.

From a user point of view, obtaining a structural model of a protein from its primary sequence is very simple. You just submit a sequence to a 3D prediction server and get a PDB file with the predicted structure. The difficult part is not getting the structure, but knowing to what extend you can rely on it. There are some *a priori* checks to assess how good the prediction will be and some *a posteriori* checks once you obtain a model.

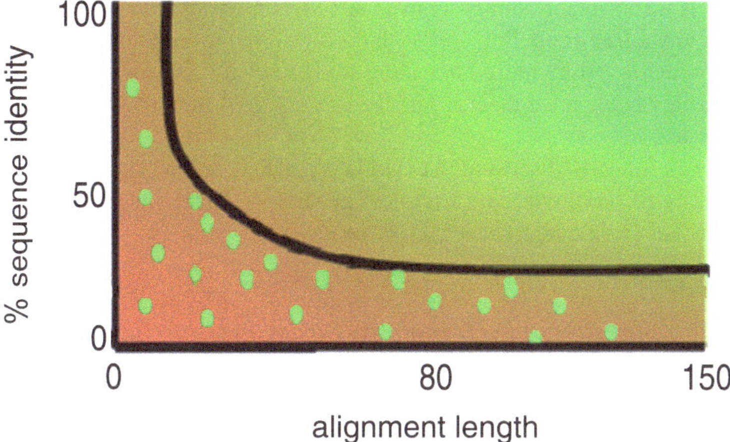

Fig. 2.13 Proportion of pairs proteins with similar 3D structures depending on the sequence identity and the length of the aligned region. Adapted from (Sander and Schneider 1991). The proportion of pairs with similar structure is indicated by a color scale, from *green* (many pairs with the same structure) to *red* (no pairs with the same structure). The *green dots* try to represent cases of pairs with similar structure in spite of undetectable sequence similarity

The most important *a priori* check is the similarity, and therefore the conservation, between your protein sequence (known as "target") and the sequence of the homologous protein whose structure has been solved. This known structure is called "template". The approach that builds a 3D model for a target based on a homologous template is called **template-based modeling**. In different studies based on proteins of known structure people quantified to which extend pairs of proteins with a given percentage of sequence identity have the same 3D structure or not. The results could be summarized in Fig. 2.13, adapted from (Sander and Schneider 1991). It can be seen that almost all pairs of proteins with sequence identities of 25 % or higher (over an alignment region longer than 70 residues) have virtually the same 3D structure. For shorter aligned regions, a larger sequence ideintity is required for being sure that the structure is going to be the same. These data provide a sort of upper limit for the results we can expect from template-based modeling. The cases of similar structures in spite of low sequence identities (green dots in Fig. 2.13) represent non-homologous proteins with the same fold. It is difficult to take advantage of those for predicting structure by template-based modeling since these cases can not be detected by plain sequence searches. The approach which explores these cases for predicting structure, known as "fold recognition" is not covered in this book. Nevertheless, the "fragment-based modeling" commented latter could pick up some of these relationships.

So the very first step in modeling a protein structure is to check whether there is a suitable template (i.e. a solved protein structure in PDB) in order to do a template-based model. Most programs are able to automatically discover suitable templates (running in automatic mode) but you can also "force" them to use a particular template.

If there is no detectable similarity to an already known structure, then you would need to follow a **free-modeling** approach (also referred to as *de-novo* or *ab-initio*). These approaches, while more interesting from a scientific point of view, are not covered in this book since their accuracies are not very high and they are usually not easy to use.

A posteriori checks, discussed in detail later, include a wide variety of estimates of the quality of the model you have just generated. These can be of two types: global (for the whole model) or local (per residue).

2.5.1 Template-Based (Homology-Based Approaches)

Swiss-Model (Schwede et al. 2003) is the server to use when there is a good sequence identity between your protein sequence (target) and the template uniformly distributed along both sequences, that is, when that single template can be used to model the whole length of the target. You can a test that by yourself (running a blast against PDB and checking the alignments) or run Swiss-Model right away since that search for a suitable template is the first thing the server will do.

Swiss-Model—Prediction of protein structure (homology-based approach)	http://swissmodel.expasy.org/ http://csbg.cnb.csic.es/PB/T1220	

In the main form click the "Start Modelling" button.

Paste your protein sequence (the target sequence) or upload a sequence file, provide a name for this job (or project) and optionally an e-mail address to receive the results. At this stage you can click on "Search For Templates" to have some control on the template selection; or "Build Model" to totally rely on Swiss-Model for the template selection and generate the model structure. If you decide to "Search For Templates", you can choose which template(s) to use among a list of available structures found by Swiss-Model in the PDB. In general, start with the "Build Model" option and, if in forthcoming steps (including the quality checks) you find problems apparently due to the template selection, try the manual "Search for templates".

When the process is completed, an interactive page with the model(s) generated shows up. The server generates a limited number of alternative models for you to choose.

For each model, you have a panel with the following items (Fig. 2.14): on the left there is an image with its structure, below which there is a blue bar representing the region of the target sequence covered by the model (with the mouse over it, you can see the actual residue range). Below this bar, there is a link to download the model (as a PDB file) as well as other data associated to the modeling. The right

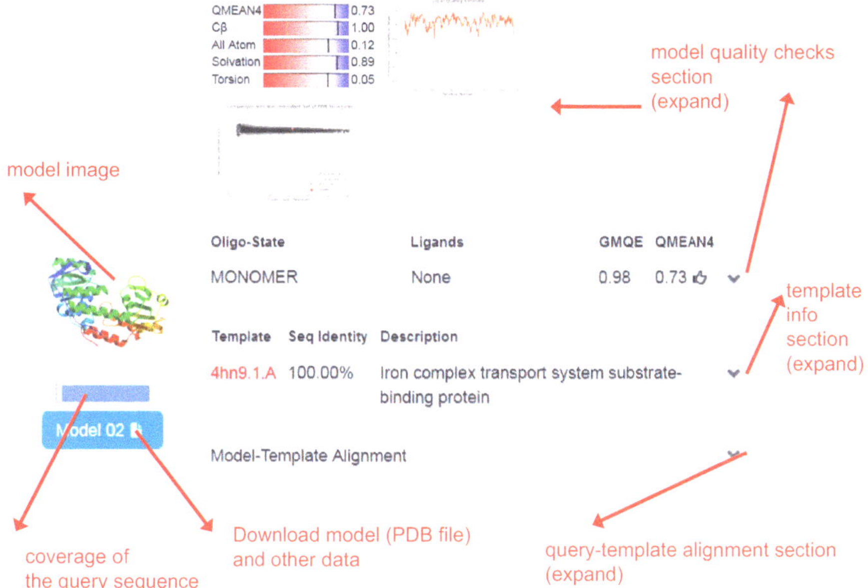

Fig. 2.14 Swiss-Model panel with model information

part of the panel contains three sections which can be expanded with the vertical arrows (Fig. 2.14). The first section contains information on the model quality. That quality is estimated using different parameters for which you can obtain detailed information on the help pages (http://swissmodel.expasy.org/docs/help). Nevertheless, the server includes convenient graphical representations which allow you to infer whether the values of these parameters for the model are "good" or not: the value for you model is compared with a distribution of values obtained for a set of real structures (Fig. 2.14) so that you can check whether your value is in a "normal range" or far apart from it. In the example in Fig. 2.14 you can see that the values of 5 parameters for that model (dark lines) are in the "good" range (blue) in the red-blue scales. The by-residue plot on the right of these scales (red line) helps inferring whether there are problematic regions in your model (low values) in spite of a globally good score, or the other way around. In the alignment section you can inspect the by-residue alignment between the target and the template. This is useful, for example, to inspect the details of this alignment for a problematic region detected in the quality plot. Indeed, by default this alignment is colored by the per-residue quality score represented in that plot, so that problematic regions are self evident (red). The tool icon close to this alignment allows changing this color schema as well as exporting the alignment in different formats.

In order to compare the different models, you can click the corresponding images and the structures will be added (structurally aligned) to the 3D viewer on the right of the page.

As a rule of thumb, a good homology model should cover a large portion of the target (unless you suspect that it is a multidomain protein and not all domains have 3D homologs), the alignment should render a percentage of identity compatible with the plot in Fig. 2.13, and present quality estimators in the normal range across its whole length. Failure of one or more of these requirements could be an indication that your sequence can not be modeled by homology based on a single template and that you should try other approaches, such as those described in the next section.

2.5.2 Template-Based (Fragment-Based Approaches)

Some template-based programs use several templates, especially in the case when no clear single template is available for the whole length of the target sequence. Partial models for different segments are then combined into the final predicted structure using *ab-initio* approaches, therefore following a fragment approach (as a kind of Dr. Frankenstein protocol). You should try one of these systems in case Swiss-Model or other systems have difficulties finding suitable templates for the whole length of the target: e.g. different independent models are generated for different parts of the target, and/or the models covering the whole length present bad quality scores. In principle, fragment-based methods are also suitable for "single-template" modeling but, since they are slightly more difficult to use and interpret the results, it is recommended to try a non-fragment method first, and turn to this category in the face of the problems commented above.

I-TASSER (Roy et al. 2010) follows this fragment-based approach, and has been consistently ranked as a top server in modeling competitions (See CASP experiments in http://www.predictioncenter.org)

I-TASSER—Prediction of protein structure (fragment-based approach)	http://zhanglab.ccmb.med.umich.edu/I-TASSER/	
	http://csbg.cnb.csic.es/PB/T1230	

Paste your protein sequence (or upload sequence file), enter your e-mail and password (you need to –freely- register previously) and click on "Run I-TASSER" to do a complete automatic structural prediction.

I-TASSER additionally allows you to set two types of options:

- Option I: to manually assign templates and or distance restraints. This option allows you to define templates to be included in the prediction. Templates can be specified without alignment (providing just PDB accession and chain identifier) or by providing target-template alignments. In the first case, I-TASSER will automatically construct the alignment between target and template sequences. In most cases, you will not need to force I-TASSER to use certain templates. But it can happen that, for example, you know by external evidences that your protein should have a certain fold and you want to force the program to use a template with it. Introducing restraints could be more useful. For example you can have external evidences that certain regions of the protein should be close to each other (e.g. disulphide bridges, active sites …). With this option, you can force I-TASSER to generate models fulfilling these restrictions.

- Option II: to exclude templates. In most cases you will not need to use this option.

I-TASSER can provide as result one or various structural models, together with an alignment of the target sequence and the templates used, and various quality parameters (including global and local values).

For each structural model, a C-Score (in the range −5 to +2) indicates the global confidence (the higher the value the better). This score is based on the quality of template alignments and the convergence parameters during the prediction process.

Only for the first model two additional global parameters are reported: estimated TM-score and estimated RMSD. These parameters are estimate values inferred from the C-Score. Generally a prediction with TM-score > 0.5 can be considered a model of correct topology, while TM-score < 0.17 indicates a poor model. The RMSD (root-mean-square deviation) estimates the average distance between the model and the "true" structure (see below), and is measured in Angstroms (therefore, the lower the better the prediction is).

In addition to these global parameters, I-TASSER provides local quality estimates for each model, defined as the distance deviation (in Angstroms) between residues in the model and the native (or "real") structure. As in the case of TM-score and RMSD, these cannot be directly measured as the native structure is not known. Therefore values are estimated from a set of parameters that can be computed like the alignment coverage, secondary structure and solvent accessibility predictions among others.

Finally, I-TASSER reports a normalized B-factor (or B-factor profile, BFP) for each residue of the target protein. Residues with BFP values higher than 2 can indicate problematic regions of the model.

2.5.3 *Model Quality Checks*

In addition to the quality measures reported by the modeling programs described previously, you can assess the quality of a model using external third-party servers. You can use them for obtaining additional quality estimators different from those reported by the prediction servers, or for evaluating generic models obtained by any other approach.

One of these applications is ***ModFOLD*** (McGuffin et al. 2013), which provides model quality parameters (both global and local) for a single model, or for multiple models.

ModFOLD—Estimation of the global and local quality of protein structural models	http://www.reading.ac.uk/bioinf/ModFOLD/ModFOLD_form_4_0.html http://csbg.cnb.csic.es/PB/T1240	

Paste the primary sequence of the target, and upload the PDB file of the model you want to assess. Finally click on "Predict".

The global quality values reported are:

- Global model quality score: ranges between 0 and 1. In general, scores less than 0.2 indicate there may be incorrectly modeled domains and scores greater than 0.4 generally indicate more complete and confident models, which are highly similar to the native structure.
- *p*-value: the fraction of models with that score that are not similar to the corresponding native structures in a test set.

The by-residue (local) quality values can pinpoint problematic regions in the model. The higher these values the worse is that part of the model predicted to be. MODFOLD generates a plot with the representation of these values along the protein sequence, as well as a 3D interactive view of the model colored by those estimators, in a color scale from blue (good –low values-) to red (bad –high values-).

2.6 Analysis of Protein Structure

Once you have the 3D structure of your protein of interest, either experimentally determined or predicted, you can extract a lot of information from it, either alone or in combination with other sources of information (e.g. multiple sequence alignments). In the case of predicted structures, you have to be aware of the expected error of your model depending on the methodology used (Sect. 2.5) and evaluate to which extent they can affect the particular tool for structure analysis you plan to use.

2.6.1 Mapping Conservation

Sequence conservation is generally used for the identification of individual residues or regions in a protein that might be functionally relevant. When the atomic structure of a protein (or a model) is known, it is interesting to map conserved as well as non-conserved regions in the protein structure. This is because the structure allows, for example, distinguishing conservation due to structural reasons (in the core of the structure) from that due to functional reasons (more in the surface). Moreover, restricting to conserved residues which are close in the 3D structure (and eventually in the surface) is a typical way of filtering a conservation-based prediction of active/functional sites.

The ***ConSurf Server*** at the Tel-Aviv University calculates the conservation scores for all residues in a protein from a multiple sequence alignment, and maps these scores (by color coding residues) in a given protein structure (Ashkenazy et al. 2010).

ConSurf—Analysis and mapping of residue conservation in protein structures	http://consurf.tau.ac.il http://csbg.cnb.csic.es/PB/T1250	

In the main web page, start by selecting 'Amino-Acids'. Then select 'YES' if there is a known protein structure, and enter its PDB ID (if protein structure is already in the PDB database) or upload your own PDB file (if you want to analyze your own structure).

Note: In the absence of a protein structure ('No') the server will switch to a different strategy: it will predict the solvent accessibility for your protein (Sect. 2.4.1) and report the conserved residues predicted as accessible.

Then select the 'Chain Identifier' in case the PDB has more than one chain.

In the next question ("Do you have a Multiple Sequence Alignment?") select 'YES' if you want to use your own MSA, and 'NO' if you want ConSurf to generate it. In case you upload a MSA, indicate the Query Sequence Name (the exact ID of the protein in the MSA file that corresponds to the PDB structure).

If you want to use your own phylogenetic tree answer 'YES' to the next question (and provide the tree file in Newick format –Sect. 1.10-). If you answer 'NO' ConSurf will calculate the tree from the MSA.

Finally choose the method and evolutionary substitution model to calculate the conservation scores (or use default options). Optionally, you can provide a job identifier, and your e-mail address to get a copy of the results.

Consurf reports the conservation in a scale which goes from "1" (blue, highly variable) to "9" (purple, highly conserved). That color schema is used for coloring the positions of the MSA as well as the residues in the 3D structure (Fig. 2.15). Surface regions of the structure enriched in conservation (purple patches) are indicative of functionally important regions.

2.6.2 Protein and Ligand Contacts

Studying the known ligands present in the structure of our protein and the residues involved in binding them, or predicting these binding sites in case we have the

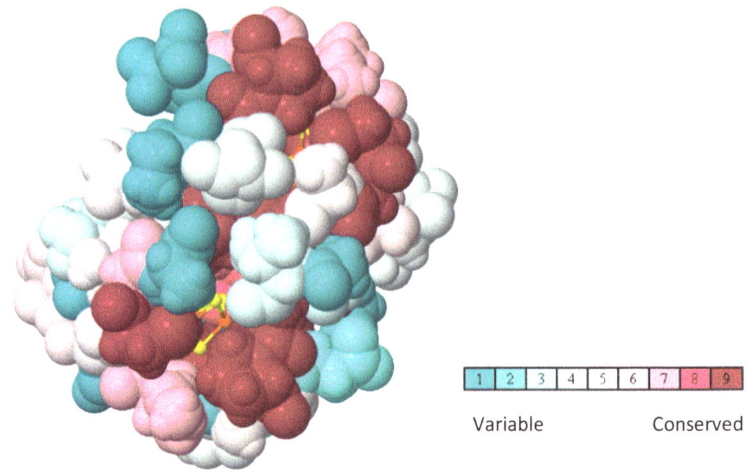

Fig. 2.15 Protein structure with residues colored according with Consurf calculated conservation

structure in *apo* (ligand-free) form, can provide valuable information on the function of our protein, as well as devise ways of designing mutants.

When a protein complex structure has been solved in *holo* (ligand-bound) form and is available on the PDB, it is a valuable source of information on the molecular interactions established between proteins, or between a protein and small ligand, or a protein and a nucleic acid (DNA or RNA). The exact residues contributing to the interaction and the nature of the physical forces established can be obtained from the **PDBsum** resource commented in Sect. 2.2. PDBSum also provides pre-calculated results of residue conservation obtained with the ConSurf Server commented above.

In case the structure we want to analyze is not deposited in PDB, we can use the ***LCP/CSU server*** at the Weizmann Institute (Sobolev et al. 1999). Among other things, this server generates a whole report, as well as graphical representations, on the contacts between the proteins and the ligands in a PDB file uploaded by the user.

LPC/CSU server—Analysis of protein-ligand interactions	http://ligin.weizmann.ac.il/cgi-bin/lpccsu/LpcCsu.cgi	
	http://csbg.cnb.csic.es/PB/T1260	

In the web form of this tool, upload the PDB file with your holoprotein(s), leave the rest of options with their defaults, and press "Run". (It is also possible to enter a PDBID in case the structure is deposited).

In case your structure contains more than one ligand, the next screen allows us to select that you are interested in. In the results window, there is a JMol applet where, by default, the ligand and the residues contacting it are represented. On the left, you have a sort of menu with links that take to detailed tables and pages with information on the individual residues/atoms involved in the interactions. These detailed data are shown in the lower panel as we click the menu items.

2.6.3 *Surface Clefts, Binding Pockets, Tunnels and Internal Cavities*

These types of topological features of the protein structure are in many cases indicative of active sites or functional regions. Consequently, it might be interesting to locate them in our structure as a complement to the other methods for predicting functional sites discussed.

For structures already deposited in the PDB/RCSB, the ***PDBSum*** resource commented above (Sect. 2.2) contains their pre-calculated clefts and tunnels. These are available in the "clefts" and "tunnels" tabs of the entry pages. By default, these tabs contain a list of the clefts/tunnels detected in the structure, identified with different colors. They are highlighted in the structure of the protein and their characteristics (size, radius, depth, length, …) are shown in the list.

If you have a structure not deposited in PDB, you can use the ***IsoCleft Finder*** at the EBI (Najmanovich et al. 2008). This server not only detects the clefts on an uploaded structure, but it also compares these found clefts with a database of clefts found in other structures associated to functional sites. This is a way of predicting function and functional sites based on similarity of clefts (binding pockets, active sites, …)

IsoCleft Finder—Detection of clefts in structures and search for structures with similar clefts	http://www.ebi.ac.uk/thornton-srv/databases/cgi-bin/icfdb/StartPage.pl http://csbg.cnb.csic.es/PB/T1270	

For analizing your structure, upload the corresponding PDB file in the section "B." of the form and press "Submit". If you have not provided the optional email for being informed of job completion, a dialog box shows up for confirming that you will wait in front of the browser. The process of cleft finding is quite fast and hence in most cases the email confirmation is not necessary. Nevertheless, if you plan to do a forthcoming search with the found clefts (see below) it is highly recommended to include an email since this second part can really take a long time.

The results page contains a list of the clefts found in the structure. For each cleft, you can find a JMol applet with a representation of that cleft (residues defining it and, eventually, the ligand within it) as well as a list of the residues. Each cleft has a checkbox to decide whether you want to include it in the forthcoming database search for similar clefts or not. The individual residues of the clefts can also be checked/unchecked so as to restrict the search to parts of the clefts. Once you have selected the clefts and residues you want to use for the search (all by default), press "Submit".

The result page with the similar clefts found contains the following information. For each cleft of your structure, a table with a list of similar clefts found in PDB is shown, with Z-scores and P-values as indicators of the confidence of the similarity match. The PDB IDs are active links to the corresponding pages in the PDBSum resource (Sect. 2.2) where detailed information on that structure and its bound ligand(s) can be found. For each "similar" cleft, the PDB code of the structure containing it, as well as the bound ligand (in PDB 3-letter nomenclature) are indicated. For each cleft, there is also a matrix showing the pairwise similarities of the ligands listed in the table. This ligand similarity is associated to a colour scale, from blue (high similarity) to white (low similarity). This matrix can be used to infer the "homogeneity" of the set of ligands found in the similar clefts.

A similar service is provided by the ***CASTp server*** at the University of Illinois (Dundas et al. 2006), which also detects internal cavities.

CASTp—Identification of surface pockets and internal cavities	http://sts.bioengr.uic.edu/castp/calculation.php http://csbg.cnb.csic.es/PB/T1280	

Just upload the local PDB file to analyze and click the "Submit" button.

The results page contains on the left a list of the clefts found indicating their volume and area. Selecting one or more they become highlighted (in different colors) in the JMol representation of the structure (right) and on the sequence of the protein below it.

The ***MOLEonline*** server at the Univerzita Palackého v Olomouci allows calculating tunnels, pores and cavities in protein structures (Berka et al. 2012).

MOLEonline 2.0—Identification of tunnels, pores and cavities in structures	http://mole.upol.cz/online/ http://csbg.cnb.csic.es/PB/T1290	

For starting the calculation, upload the file with your structure (PDB format) or enter the PDB ID code for deposited structures and press "Next". In the next page select "Automatic starting points" and press "Submit".

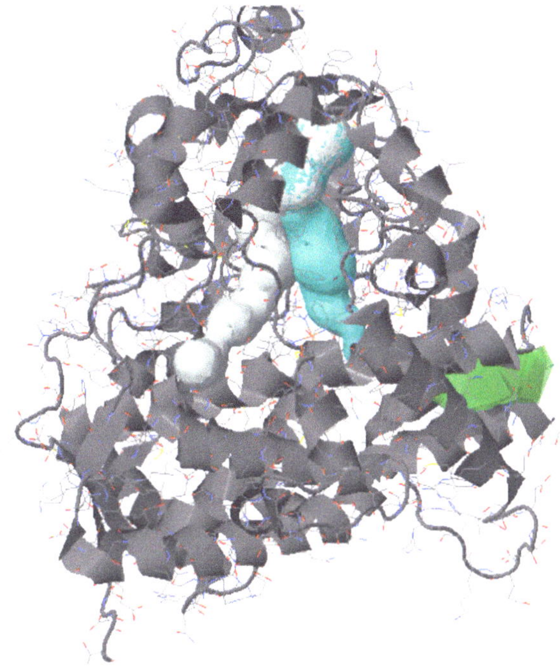

Fig. 2.16 Some of the tunnels and surface clefts found by MOLEonline in the structure of the human microsomal cytochrome P450 3A4 (PDB ID: 1TQN)

In the results page, a JMol applet with the structure is shown. On the right, there are panels for the different features (tunnels, cavities …) which can be expanded/collapsed with the "+"/"-" buttons depending on those you are interested in. Basically these panels list the instances of that type found, including information on their characteristics and links to extended information (residues delimiting them, etc.). Each feature has a checkbox to show/hide it in the JMol representation (Fig. 2.16).

Chapter 3
Systems

3.1 Introduction

In spite of the wealth of information we can infer from the analysis of a protein sequence or structure, we know that proteins do not work alone within the cell. Recent advances in experimental techniques allow us to study whole biological systems at the molecular level, obtaining results involving hundreds or even thousands of protein/genes in a single shot. When you perform such a large-scale experiment, you will probably need to understand and interpret those results (basically long list of genes/proteins) in biological terms. Therefore, the focus of the tools described in this chapter is not to study in detail individual proteins, but to obtain information from large protein sets and the different networks of relationships among them.

3.1.1 *Protein/Gene Functional Annotations*

Many of the tools commented in this section work with protein functional annotations. They basically mine the annotations associated to a large set of protein/genes coming from a large-scale experiment and try to infer functional information from it.

Basically a functional annotation is the association to a protein of one or more terms of a defined vocabulary. The vocabularies of the first functional annotations contained a very limited number of terms, for example a dozen of broad functional classes such as "metabolism", "transmission of genetic information", … The de-facto standard today for functionally annotating proteins is that defined by the ***Gene Ontology (GO)*** consortium (Harris et al. 2004). Since it is used by most tools and its structure is a little bit complex, it is worth including here a brief description, for you to know what you are handling when you work with GO functional annotations.

In short, GO defines three vocabularies to represent three different aspects of the complex phenomenon of protein function: "molecular function" (MF), "cellular compartment" (CC) and "biological process" (BP). Within each vocabulary, the terms are not at the same level but related in a complex network of parenthood relationships, so that a given term is related with a number of parental terms via

F. Pazos, M. Chagoyen, *Practical Protein Bioinformatics*,
DOI 10.1007/978-3-319-12727-9_3

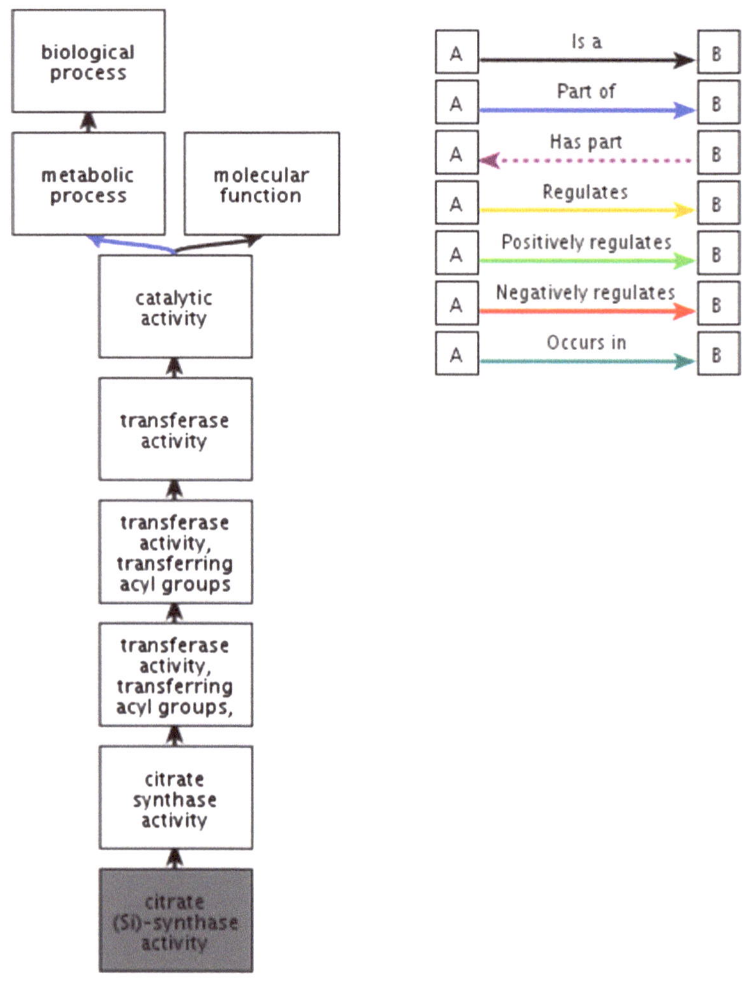

Fig. 3.1 Pathway of parenthood relationships for GO term "citrate synthase activity" (GO:0004108). (Obtained from the GeneOntology web site (www.geneontology.org))

relationships such as "is_a" or "part_of". Consequently, the terms of this network represent functional aspects of different specificity, from very broad aspects (e.g. "catalytic activity") to very specific ones (e.g. "citrate synthase activity"), in such a way that broad terms are parents of more specific ones. For example, Fig. 3.1 shows the ancestors of the "citrate synthase activity" term (GO:0004108), up to the most general "biological process" term.

A protein (gene product) becomes annotated by assigning to it one or more GO terms from the three categories. Obviously, assigning a term implicitly involves assigning also all its ancestors.

Apart from being used in many databases and tools, GO has its own web site where you can directly explore/navigate the ontology structures, retrieve gene/protein annotations, as well as obtain more information on this project.

Gene Ontology—Structured representation of molecular biology knowledge, and annotations of gene products	http://geneontology.org/ http://csbg.cnb.csic.es/PB/Y1010	

3.1.2 *ID Conversions*

Depending on the experiment you have performed, and more generally, on the bioinformatics tools you have used to perform the primary analysis of your data, you will end up with protein/gene lists given in terms of identifiers from different databases. These identifiers might refer to gene nomenclature (hopefully official gene symbols), to DNA/RNA sequence collections (e.g. GenBank), to protein sequences (e.g. UniProt) or to general sequence collections (e.g. UniGene), among others. Since the analysis tools commented below might need your list of genes/proteins with identifiers of a particular database, there is a tool which can "translate" a long list of identifiers of a given database to the corresponding ones in other resources.

The ***UniProt*** site (Sect. 1.3) provides a convenient tool to convert a list of IDs to/from UniProt accessions numbers (AC) or entry names (ID). You can access it following the "Uploadlists" tab from the main menu. Paste the list of identifiers (or upload a text file containing them), select the type of input identifiers and the output identifiers you want to convert to. Identifiers supported cover a very wide range of databases, but conversions can only be done to or from UniProt identifiers (AC or ID).

3.2 Annotation Enrichment Analysis of Large Proteins Sets

Enrichment tools (Huang et al. 2009) are able to analyze the functional annotations of all proteins in a set, and extract those annotations particularly enriched in the input set respect to a background set. These tools provide statistical estimates (p-value and others) associated to the annotations measuring the significance of finding that particular annotation in a set of such size just by chance. Enrichment analysis is the most common way of interpreting a large set of genes/proteins in functional terms. Because these tools were initially created for the interpretation of DNA microarray experiments, most of them are designed to work with genes. Nevertheless, from a practical point of view, they can be used in the analysis of protein sets, as they analyze the functions of the gene products, instead of the genes.

For obtaining a statistical estimate, it is necessary to compare the annotations in the set of proteins you are interested in with those in a background set. The exact background to be used in the analysis depends on the experiment you have performed. For example, in a proteomics experiment where 'ideally' all proteins in the proteome could be identified, the list of proteins showing up in the experiment should be compared to the full proteome (that is, all proteins known for a given organism). In a "closed" experiment, like a DNA microarray, the background could be the set of probes in a particular chip.

DAVID (Database for Annotation, Visualization and Integrated Discovery) (Dennis et al. 2003) integrates functional annotations from different resources and allows analyzing datasets from a diverse number of organisms. It integrates more than 40 types of annotations, including Gene Ontology terms, protein domains, pathways and diseases.

DAVID gene—Functional (enrichment) analysis of a set of proteins	http://david.abcc.ncifcrf.gov/ http://csbg.cnb.csic.es/PB/Y1020	

To perform an analysis, select "Functional annotation" at the left panel (labeled "Shortcut to DAVID Tools"). Then, select the "Upload" tab from left panel. Paste a list of IDs or upload the file containing them. Select the type of ID from the menu provided ("Step 2: Select Identifier") and choose "Gene List" from "Step 3: Select Identifier". Then click the "Submit List" button. DAVID will now show you the list of species that matches your list of IDs. Select the species you are interested in, and click the "Select Species" button. You have defined the set to analyze.

Optionally, you can now define a background set for statistics calculation. To choose among pre-defined backgrounds click the "Background" tab on the left panel. Alternatively, enter your own background set by uploading a second list: go to the "Upload" tab from left panel, follow steps 1 and 2 as previously, and choose "Background" in "Step 3: List Type". Click on "Submit List" to send the background data.

Now, you are ready to perform analysis. You can use the default annotations or, optionally, mark those you are interested in. In this case expand the annotation categories by clicking the "+" symbols and select/deselect annotations. DAVID provides three types of analysis:

1. Click on "Functional Annotation Chart" to display a table with every individual annotation found within the set and its corresponding statistical estimates: p-value and multiple testing corrections (e.g. Benjamini).

Example: "Annotation chart" or annotation view.

Category	Term	Genes	Count	%	P-value	Benjamini
GOTERM_BP_FAT	Forebrain development	…	17	34.0	6.9E-20	1.1E-16
SP_PIR_KEY-WORDS	Proto-oncogene	…	16	32.0	9.9E-18	1.6E-15

2. Click on "Functional Annotation Clustering" to display groups (clusters) of related annotations. DAVID assumes that annotations are related if they are associated to the same gene products. The same statistical estimates as those reported in the Annotation Chart are shown.

Example: "Annotation clustering" or grouped annotation view.

Category	Term	Genes	Count	P-value	Benjamini
Cluster 1					
GOTERM_BP_FAT	cell motion	…	17	2.8E-18	2.3E-15
GOTERM_BP_FAT	cell migration	…	18	3.9E-17	1.3E-14
…	…	…	…	…	…
Cluster 2					
GOTERM_BP_FAT	response to radiation	…	12	1.1E-10	3.4-E-9
GOTERM_BP_FAT	cell death	…	14	1.1E-6	1.6E-5

3. Click on "Functional Annotation Table" to show all the annotations found for each gene product in the list. No statistical estimates are provided.

Example: "Annotation table" or gene/protein view:

ID	Gene name	GOTERM_BP_FAT	KEGG_PATHWAY
P26358	DNA (cytosine-5-)-methyltransferase 1	GO:0006306~DNA methylation,…	hsa00270:Cysteine and methionine metabolism
P01023	alpha-2-macroglobulin	GO:0002526~acute inflammatory response, …	hsa04610:Complement and coagulation cascades
…	…	…	…

In all the three results views (Chart, Clustering or Table) you can download the results for further use. Follow the "Download file" link on top of the table (marked with a Disk icon).

3.3 Protein Interaction Networks

Studying the context of our protein(s) of interest in an interaction network can provide clues about its function. If the protein is of unknown function, the biological process where it is involved can be inferred from those of its interactors ("guilty by association"). For a protein of known function, knowing its interactors (either predicted or experimentally determined) allows connecting it to other cellular processes and devising new possible roles for it. Mapping a large set of proteins in the interactome can also serve as a way of evaluating their functional homogeneity, detecting outliers, inferring the biological process enriched on them, etc. (see Sect. 3.2 above).

The web resource with the best equilibrium between information on protein interactions and ease to use is the ***STRING server*** at the EMBL (von Mering et al. 2003). STRING stores, in a common framework, massive information on experimentally determined as well as predicted interactions and functional associations for proteins. STRING is the easiest and fastest way of obtaining a summary of all available interaction information for our protein(s) of interest.

The different sources of information related to interaction are called "evidences" in STRING. For a given pair of proteins, the experimental evidences include its presence in the databases compiling the results of high- and low-throughput determinations of interactions. The prediction evidences include sequence and genomic-based features, such as co-presence in genomes ("phylogenetic profiling"), or conservation of genomic neighboring. These computational methodologies (also termed "genomic context based methods") (Harrington et al. 2008) are indicators not only of physical interaction but of functional relationships as well. Co-expression of the two genes is another evidence of interaction, as well as co-mentioning in the scientific literature ("text mining"). Finally, another evidence of possible interaction is the fact that orthologs of the two proteins in another organism have been reported to interact ("homology"). Since these evidences reflect not only direct physical interactions but also indirect relationships, we can say that the goal of STRING is to store functional associations, understood in the broadest possible sense.

STRING provides information on all the proteins inferred to interact (functionally or physically) with our sequence of interest based on one or more of these diverse evidences. For each pair, a numerical score which tries to summarize all the evidences is also provided. This score was "trained" with a golden set of interacting pairs so that it gives an idea on the reliability of the interactions.

STRING—Protein interactions and functional relationships inferred from diverse evidences	http://string-db.org/ http://csbg.cnb.csic.es/PB/Y1030	

The input for STRING is a single protein, or a set of proteins. These can be entered either by their names or IDs in different databases, or by their sequences, with the "search by name" (or "multiple names") and "search by protein sequence" (or "multiple se-

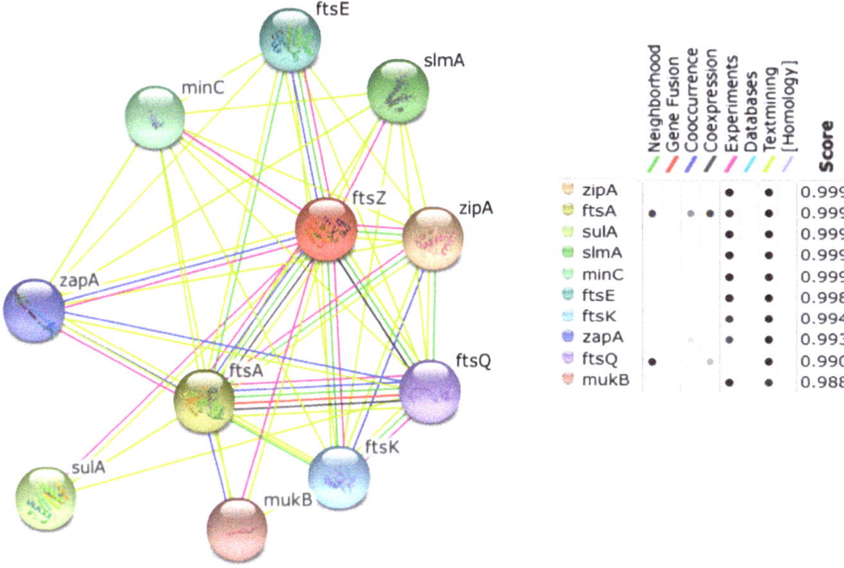

Fig. 3.2 Network of interations and functional associations around the *E. coli* FtsZ protein as stored in STRING. The table on the *right* contains the list of interactors, with a summary of the evidences supporting each interactions and the final STRING score

quences") tabs of the web form respectively. In the last case, a BLAST search is performed. If the input results in matches against several proteins in STRING's database (e.g. synonymous names, multiple BLAST matches for our sequence…), an intermediate page shows up allowing the selection of one of the choices.

When starting with a single protein, the top of the results page contains a graphical representation of the interaction network around this protein. Each node is a protein, and the connections represent different evidences of interaction, each associated to a different color (Fig. 3.2). If the 3D structure of the protein is available or has been modeled by homology, a small icon with a representation of that structure appears within the corresponding node.

Clicking on a node you obtain additional information on the corresponding protein, as well as links to external databases. The nodes can be moved so as to change the layout of the network. This network not only includes the inferred interactions for our protein but those among its (first level) interactors as well. This initial network contains an initial set of interactions according with a threshold of STRING's score. This set can be increased/decreased with the "+" and "–" buttons.

In principle, the more reliable interactions are those associated to a large number of evidences (colored lines) and to high scores. The "confidence" button changes the representation so that the lines connecting the nodes are not colored by evidence but their width become proportional to the score. The "evidence" button changes the representation to the previous one (colors by evidences). Evidences based on text-mining (co-mentioning in the scientific literature) contain additional information on the type of interaction, extracted from the textual context where the two proteins are co-mentioned, and eventually its directionality (e.g. "A binds B", "A activates B", …). Such information can be displayed in the network with the "actions" button. Now the color of the lines connecting the nodes reflects these actions, and the eventual presence of arrows/dots indicate the directionality. Finally, the "save" button allows exporting this network of interactions in different formats (graphical, text, …).

Below this network representation, you find the "Predicted functional partners" table, which contains a detailed list of the predicted interactors, indicating the different evidences as well as the final score (Fig. 3.2).

The different buttons in the "View" row, one for each type of evidence, lead to specific pages with detailed information on that particular evidence for the interactions shown above, so that you can obtain additional information on a particular type of linkage. For example, for the "co-expression" evidence, you can see a square matrix of all proteins against all, where the similarity of expression profiles for each pair is indicated in a color scale. For the "text-mining" evidence, you obtain the publication abstracts where the two proteins are co-mentioned, with their names/IDs highlighted. For the genomic evidences, graphical representations of the genomic context of the genes in a number of organisms are shown so as to highlight the presence of conserved operons, gene fusion events, etc.

The "summary network" button takes you back to the original network representation of the interactions around your protein.

Searching STRING with multiple proteins allows inferring whether there are functional associations among them and how clustered they are according with STRING's evidences. Among other things, this can serve as a way of analyzing the functional homogeneity of this set.

3.4 Metabolic Networks

If your protein(s) of interest are enzymes, there are many metabolic-related resources where you can find information on them. Although, in general, these resources are not centered in proteins, there are still many ways in which they can be queried from a "protein-centric" point of view.

3.4.1 Retrieve the Metabolic-Related Information Associated to a Protein of Interest

The largest resource with metabolic information is the ***"Kyoto Encyclopedia of Genes and Genomes" (KEGG)*** (Kanehisa et al. 2004). The aim of KEGG is to store all information related to metabolism (metabolites, enzymes, genes, pathways …) in an integrated way so that it can be queried and navigated. In the following, we will only describe the KEGG features more directly related to proteins (enzymes).

KEGG—Integration of chemical, genomic, systemic and functional biological information	http://www.genome.jp/kegg/kegg2.html http://csbg.cnb.csic.es/PB/Y1040	

The easiest way to locate your protein(s) of interest in KEGG is the BLAST search option available through a number of links in the main page above. This takes you to a web form where you can introduce the sequence of your protein (or its ID in a database if you know it). Although this web form contains a lot of options, since it can be used to perform generic BLAST searches against many protein databases mirrored in KEGG, the default options/parameters are fine for searching for your protein in KEGG.

After introducing the sequence and pressing "Compute results", a typical BLAST result page shows up. Although this page contains many of the typical features present in other BLAST servers (retrieve found sequences and alignments, perform a multiple sequence alignment with the top hits, etc. –Sect. 1.6.2-), here we are only interested in locating your sequence in KEGG, typically the first hit in the list. Following that link, you end up in KEGG's information page for that gene/protein. That page contains a lot of information on that protein retrieved from other databases, as well as the corresponding links to these databases. Here we will only focus on the KEGG-related information.

There is a link to the "KEGG orthologs" group (KO) for that protein. A KO groups all homologous gene products of different organisms potentially with the same metabolic activity. Following that link you can find this list of genes ("Genes" section) as well as their taxonomic distribution ("Taxonomy" button). This button takes us to a taxonomic tree in which the organisms containing an ortholog of your protein (hence potentially presenting that enzymatic activity) are colored in red, and those lacking it in blue. This allows inferring to which extent the enzymatic activity of your protein is present in other organisms and which branches of the evolutionary tree they represent. The same KO section contains a link with the Enzyme commission (EC) code of the enzyme. It takes to the KEGG page for that enzyme. Among the information contained in this enzyme page, there is a list of genes coding enzymes with the same EC number in different organisms, and links to KO groups associated to that EC. Note that EC codes can be very generic and hence be associated to many different KO's without evolutionary relationship. For example EC:1.1.1.1 (alcohol dehydrogenase) is associated in KEGG to 9 different KO's representing dehydrogenases for different alcohols. This EC page also contains links to more specialized enzyme databases, for example BRENDA (Schomburg et al. 2004), which contains a lot of information on the thermodynamics and kinetic parameters of enzymes.

The "Pathways" section of the protein page contains a list of metabolic pathways where that enzyme is involved. These are links to pages with graphical representations of the generic pathways. In these representations, your enzyme is highlighted

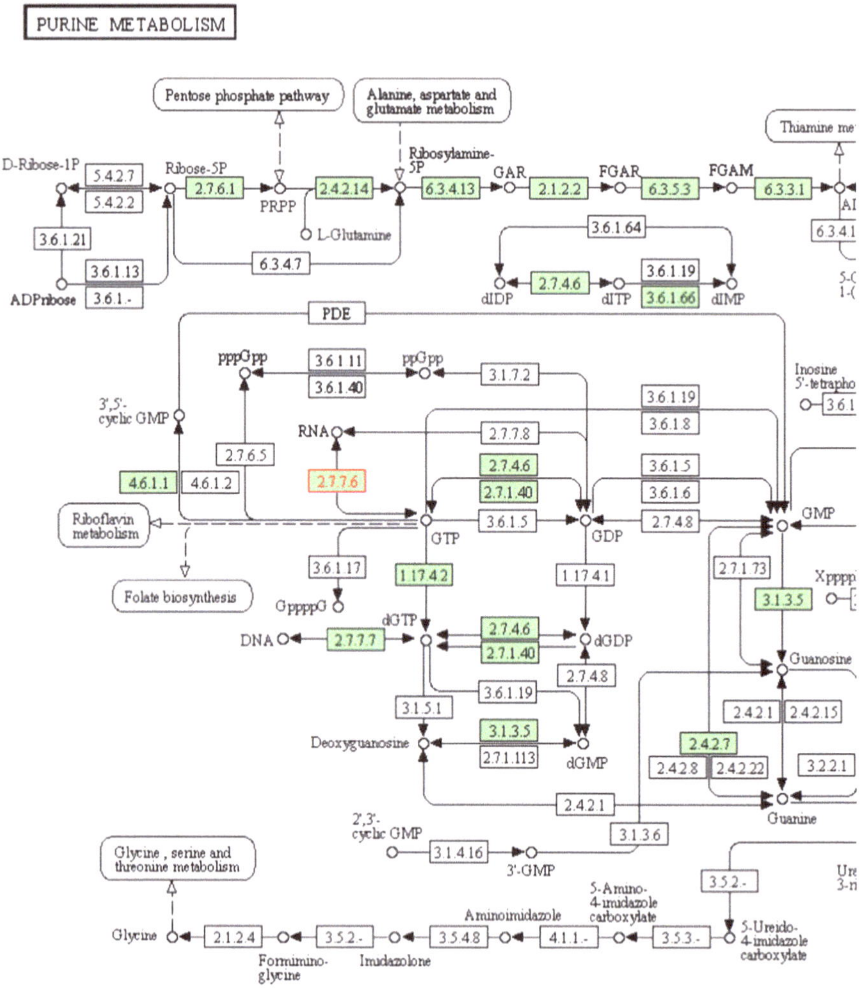

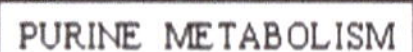

Fig. 3.3 Representation of part of a KEGG pathway, with an enzyme highlighted

in red, and the other enzymes (potentially) present in the same organism in green. Enzymatic activities not present in the organism are shown in white boxes (Fig. 3.3).

All elements in this pathway representation (chemical compounds, enzymes, entry points to other pathways …) are active links to the corresponding KEGG pages.

The "KEGG compounds" link on the "All links" section on the right of the protein page takes to a page with a list of chemical compounds associated to that enzyme (either as substrates, products or cofactors) which in turn are active links to the corresponding pages in KEGG with extensive information on these metabolites.

Another interesting link of the "All links" section is "KEGG reactions". It takes to a list of reactions potentially catalyzed by your protein, which are links to the corresponding KEGG pages with detailed information on these biochemical reactions.

3.4.2 Map a Large Set of Proteins in the Metabolome

A possible way of interpreting a large list of protein/genes in functional terms is to map them in the metabolome so as to "visualize" their metabolic context. This complements, for example, the automatic enrichment analysis commented in Sect. 3.2. Mapping a large set of proteins in the metabolome allows, for example, to infer whether they belong to the same pathway (or are close in the metabolism), whether they are consecutive enzymes in a biochemical pathway (eventually with "gaps" –protein not in our list- which can point to further experiments), etc.

The ***KEGG*** resource, commented above, includes a "mapping" feature for coloring different KEGG objects (chemical compounds, enzymes, reactions…) in the graphical representations of the metabolic pathways. You can make use of this feature to map a set of proteins. Within KEGG's main page, follow the "KEGG Mapper" link, and then "Search&Color Pathway" on the left. In the web form, you have to first enter the organism in whose metabolism you want to map your list of proteins ("Search against" box). KEGG uses a three-letter code for indentifying the organisms (e.g "hsa" for human; "eco" for E coli K-12 …). If you know the code for your organism, enter it directly in the box. If not, press the "org" button besides it and search for the organism in the provided form. In "Primary ID", select the database to which the IDs of your list belong to (e.g. "Uniprot"). Finally, paste in the "Enter objects…" text box the list of IDs of your proteins (one per line), as in the following example:

```
P53396
P21399
Q9HBB2
V9HWB7
O75874
V9HWJ2
P48735
B4DFL2
B4DSZ6
```

Optionally, you can specify the foreground and background colors that will be used for highlighting these items in the pathway representations:

```
P53396 red, white
P21399 brown, white
Q9HBB2 brown
V9HWB7 yellow
O75874 yellow
V9HWJ2 yellow, black
P48735 red
B4DFL2 blue
B4DSZ6 blue
```

After entering this list in the box, or uploading it from a file with the "Browse" button, you can press "Exec".

You are taken to a page with the list of KEGG pathways containing one or more of your proteins. The number of proteins within each pathway is indicated between brackets. The list is sorted by this number so that the pathways at the top are those "enriched" in our proteins of interest, what provides a first idea of where our proteins are and to which extent they are "clustered" in the metabolism. When interpreting this list, it is important to take into account that KEGG pathways are hierarchical (some pathways are included into others) and that there are intersections between them. Indeed, one of the pathways within this list will be "metabolic pathways", which contains the whole metabolome of the organism. Clicking on the items of this list you obtain a graphical representation of the corresponding metabolic pathways (or the whole metabolic network) with your proteins (enzymes) highlighted.

The "interactive pathway explorer" at the EMBL (***iPath***) (Yamada et al. 2011) also allows highlighting a set of user-provided objects in representations of the metabolic network. Its interface and general "philosophy" is quite similar to the iTol system for representing phylogenetic trees commented in Sect. 1.10.

iPath—Interactive visualization and manipulation of metabolic maps	http://pathways.embl.de/ http://csbg.cnb.csic.es/PB/Y1050	

iPath imports the metabolic information from KEGG. The main differences with that resource are: (i) in addition to KEGG pathways, it contains a number of curated organism-specific metabolic networks (while the metabolic networks in KEGG are predicted from genomic information); and (ii) its interactivity and graphical capabilities are higher.

As with KEGG, we are only addressing here the more "protein-centric" features of iPath, and it is not our intention to provide a whole description of the capabilities of the system.

The main entry point to iPath is within the "iPath v2: the main interface" section of the main page (click on the image). This takes you to a representation of the complete (generic) metabolism, where the main pathways are highlighted in different colors. You can zoom in/out this representation with the mouse wheel or the controls in the floating grey toolbar (initially on the top-left corner). It is also possible to move through the representation by dragging with the left mouse button or with the arrow controls of the toolbar. When the zoom is large enough, the names of the subpathways as well as circles representing the individual metabolites appear. By clicking an item in this representation (chemical compound –circle- or chemical reaction –line-) a floating window shows up with detailed information on that item and links to external databases.

The "Customize" button on the top-right corner allows mapping a generic set of objects, including proteins, so as to highlight them in the metabolic representation. Within this form you have to provide your list of proteins in the "Element selection" box (or upload it from a file), and optionally give a name to your selection ("Selection name"). A number of IDs from different databases are allowed (see the "Help" section for a whole description). For the same example used above with KEGG (Uniprot IDs), the list of proteins would have to be provided in the following format:

```
UNIPROT:P53396
UNIPROT:P21399
UNIPROT:Q9HBB2
UNIPROT:V9HWB7
UNIPROT:O75874
UNIPROT:V9HWJ2
UNIPROT:P48735
UNIPROT:B4DFL2
UNIPROT:B4DSZ6
```

As with KEGG, you can include additional parameters to control the representation of your proteins, such as the color (in hex-RGB format, see Sect. 1.10), the width of the lines and their opacity. For example, the following will represent the first four proteins (reactions) in blue, with a line-width of 20 pixels, and half-transparent:

```
UNIPROT:P53396 #0000ff W20 0.5
UNIPROT:P21399 #0000ff W20 0.5
UNIPROT:Q9HBB2 #0000ff W20 0.5
UNIPROT:V9HWB7 #0000ff W20 0.5
UNIPROT:O75874
UNIPROT:V9HWJ2
UNIPROT:P48735
UNIPROT:B4DFL2
UNIPROT:B4DSZ6
```

Take a look at the Help section for a whole description of the format.

The rest of the controls within this form are used to specify the default representation (that for the items not included in your selection). By default, they are colored grey and drawn with thin lines to better highlight the selection. It is also possible to keep the original, pathway-based, colors for these items with the checkbox provided.

Pressing "Submit data and customize maps" updates the metabolic representation according with the information we have provided in this form (Fig. 3.4).

Finally, the "Export" button allows exporting this modified representation (eventually with your items highlighted) to different graphical formats.

3.5 Other Biological Networks

There is an increase interest in approaching biological phenomena from a network point of view. Unfortunately, the research in this area is slowly reaching the stage at which the methods/results are implemented in web servers so as to be used by

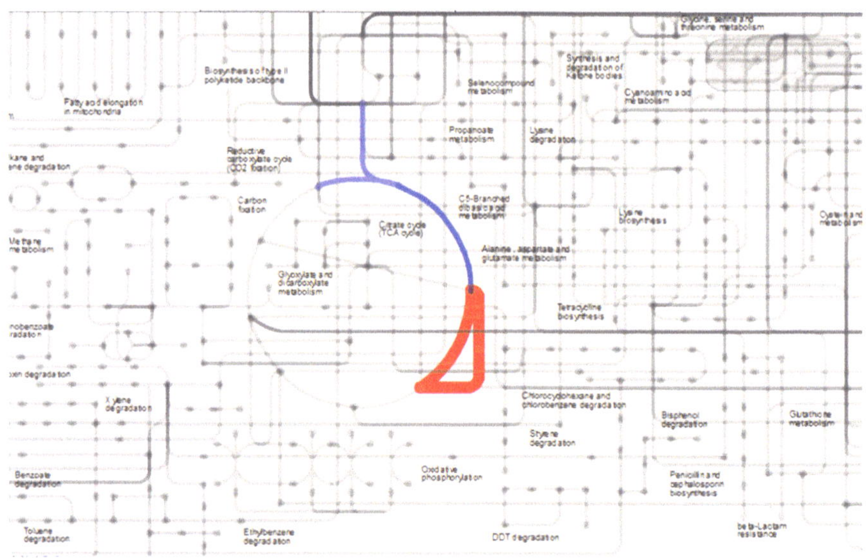

Fig. 3.4 iPath metabolic representation with some enzymes (reactions) highlighted

the community, except for the main molecular networks (metabolome and interactome, previous sections). Moreover, as happened with the visualization of protein 3D structures, the best software tools within this area comprise stand-alone programs which require local installations, configuration, etc., which are not covered in this book. Again, we recommend readers particularly interested in this subject to check these advanced software solutions for network visualization/analysis, such as Cytoscape (Cline et al. 2007) (http://www.cytoscape.org/).

Other important type of biological networks are the signaling pathways. Signaling pathways have been extensively studied due to their relationship with diseases such as cancer. Part of the information of signaling cascades should be incorporated in the protein interaction resources, since in these pathways the transmission of signals is mediated by (transient) protein interactions. Moreover, some information on signaling pathways has been incorporated in KEGG and, consequently, in the resources fed from it. Nevertheless, there are other resources with a stronger emphasis on this type of networks.

One of such resources is ***Reactome*** (Croft et al. 2014), a curated human pathway database. Pathways for other 20 species (including mouse, rat, chicken, worm, fly, yeast and plants) are inferred from human data through the recognition of orthologous proteins. Pathways in Reactome include metabolism, signaling cascades, immune system, transcriptional regulation, apoptosis and disease-related networks.

Reactome allows not only to map/visualize a set of proteins within its pathways, but also to perform enrichment analysis. Another advantage is that the layout of the elements in their pathway representations resembles those manually generated for textbooks and similar.

Reactome—Pathway mapping and analysis of a set of proteins. Focused on signaling pathways	http://www.reactome.org http://csbg.cnb.csic.es/PB/Y1060	

The main entry points to Reactome are "Browse pathways" and "Analyze data" buttons of the main page. The pathways browser shows on the left the list of pathways, in a hierarchical structure whose nodes can be expanded/collapsed. At the top of the page, you can select the organism you are interested in. Once a pathway is selected in the list, its graphical representation is shown on the right. You can zoom in/out and move with the controls on the top-left corner. Since Reactome contains different types of pathways, it has a complex set of symbols to represent the different set of molecules (proteins, metabolites) as well as "relationships" (physical interaction, regulation, metabolic transformation,…). The legend explaining the symbols is available on the blue arrow icon (top-right). Below the pathway representation there is a data panel with different tabs for obtaining detailed information on the items associated to that pathway.

The "Analyze data" button of the main page allows uploading files with your data (protein sets, expression values …) which will be incorporated on the pathway representation described above.

Bibliography

Altschul SF, Gish W, Miller W, Myers EW, Lipman DJ (1990) Basic local alignment search tool. J Mol Biol 215:403–410

Altschul SF, Madden TL, Schaffer AA, Zhang J, Zhang Z, Miller W, Lipman DJ (1997) Gapped BLAST and PSI-BLAST: a new generation of protein database search programs. Nucleic Acids Res 25:3389–3402

Andreeva A, Howorth D, Brenner SE, Hubbard TJ, Chothia C, Murzin AG (2004) SCOP database in 2004: refinements integrate structure and sequence family data. Nucleic Acids Res 32(Database issue):D226–D229

Armougom F, Moretti S, Poirot O, Audic S, Dumas P, Schaeli B, Keduas V, Notredame C (2006) Expresso: automatic incorporation of structural information in multiple sequence alignments using 3D-Coffee. Nucleic Acids Res 34(Web Server issue):W604–608. doi:10.1093/nar/gkl092

Ashburner M, Ball CA, Blake JA, Botstein D, Butler H, Cherry JM, Davis AP, Dolinski K, Dwight SS, Eppig JT, Harris MA, Hill DP, Issel-Tarver L, Kasarskis A, Lewis S, Matese JC, Richardson JE, Ringwald M, Rubin GM, Sherlock G (2000) Gene ontology: tool for the unification of biology. The gene ontology consortium. Nat Genet 25(1):25–29

Ashkenazy H, Erez E, Martz E, Pupko T, Ben-Tal N (2010) ConSurf 2010: calculating evolutionary conservation in sequence and structure of proteins and nucleic acids. Nucleic Acids Res 38:W529–W533

Berka K, Hanák O, Sehnal D, Banás P, Navrátilová V, Jaiswal D, Ionescu CM, Svobodová VR, Koca J, Otyepka M (2012) MOLEonline 2.0: interactive web-based analysis of biomacromolecular channels. Nucleic Acids Res 40(W1):W222–W227

Berman HM, Westbrook J, Feng Z, Gilliland G, Bhat TN, Weissig H, Shindyalov IN, Bourne PE (2000) The protein data bank. Nucleic Acids Res 28:235–242

Cline MS, Smoot M, Cerami E, Kuchinsky A, Landys N, Workman C, Christmas R, Avila-Campilo I, Creech M, Gross B, Hanspers K, Isserlin R, Kelley R, Killcoyne S, Lotia S, Maere S, Morris J, Ono K, Pavlovic V, Pico AR, Vailaya A, Wang P-L, Adler A, Conklin BR, Hood L, Kuiper M, Sander C, Schmulevich I, Schwikowski B, Warner GJ, Ideker T, Bader GD (2007) Integration of biological networks and gene expression data using Cytoscape. Nat Protoc 2(10):2366–2382

Cole C, Barber JD, Barton GJ (2008) The Jpred 3 secondary structure prediction server. Nucleic Acids Res 36(S2):W197–W201

Croft D, Mundo AF, Haw R, Milacic M, Weiser J, Wu G, Caudy M, Garapati P, Gillespie M, Kamdar MR, Jassal B, Jupe S, Matthews L, May B, Palatnik S, Rothfels K, Shamovsky V, Song H, Williams M, Birney E, Hermjakob H, Stein L, D'Eustachio P (2014) The Reactome pathway knowledgebase. Nucleic Acids Res 42(D1):D472–D477. doi:10.1093/nar/gkt1102

Crooks GE, Hon G, Chandonia JM, Brenner SE (2004) WebLogo: a sequence logo generator. Genome Res 14(6):1188–1190. doi:10.1101/gr.849004

Chenna R, Sugawara H, Koike T, Lopez R, Gibson TJ, Higgins DG, Thompson JD (2003) Multiple sequence alignment with the Clustal series of programs. Nucleic Acids Res 31(13):3497–3500

F. Pazos, M. Chagoyen, *Practical Protein Bioinformatics*,
DOI 10.1007/978-3-319-12727-9

de Lima MDA, Fang H, Rackham OJ, Wilson D, Pethica R, Chothia C, Gough J (2011) SUPERFAMILY 1.75 including a domain-centric gene ontology method. Nucleic Acids Res 39(Database issue):D427–D434

Dennis G, Sherman BT, Hosack DA, Yang J, Baseler MW, Lane HC, Lempicki RA (2003) DAVID: database for annotation, visualization, and integrated discovery. Genome Biol 4(5):P3

Dereeper A, Guignon V, Blanc G, Audic S, Buffet S, Chevenet F, Dufayard J-F, Guindon S, Lefort V, Lescot M, Claverie J-M, Gascuel O (2008) Phylogeny.fr: robust phylogenetic analysis for the non-specialist. Nucleic Acids Res 36:W465–W469

Dosztanyi Z, Csizmok V, Tompa P, Simon I (2005) IUPred: web server for the prediction of intrinsically unstructured regions of proteins based on estimated energy content. Bioinformatics 21(16):3433–3434

Dosztanyi Z, Meszaros B, Simon I (2009) ANCHOR: web server for predicting protein binding regions in disordered proteins. Bioinformatics 25(20):2745–2746

Dundas J, Ouyang Z, Tseng J, Binkowski A, Turpaz Y, Liang J (2006) CASTp: computed atas of surface topography of proteins with structural and topographical mapping of functionally annotated residues. Nucleic Acids Res 34:W116–W118

Finn RD, Clements J, Eddy SR (2011) HMMER web server: interactive sequence similarity searching. Nucleic Acids Res 39(Web Server issue):W29–37. doi:10.1093/nar/gkr367

Finn RD, Bateman A, Clements J, Coggill P, Eberhardt RY, Eddy SR, Heger A, Hetherington K, Holm L, Mistry J, Sonnhammer EL, Tate J, Punta M (2014) Pfam: the protein families database. Nucleic Acids Res 42(Database issue):D222–D230

Gasteiger E, Hoogland C, Gattiker A, Duvaud S, Wilkins MR, Appel RD, A. B (2005) Protein identification and analysis tools on the ExPASy server. In: Walker JM (ed) The proteomics protocols handbook. http://www.springer.com/life+sciences/biochemistry+%26+biophysics/book/978-1-58829-343-5. Humana Press, Springer, pp 571–607

Gille C, Fahling M, Weyand B, Wieland T, Gille A (2014) Alignment-Annotator web server: rendering and annotating sequence alignments. Nucleic Acids Res 42(Web Server issue):W3–6. doi:10.1093/nar/gku400

Harrington ED, Jensen LJ, Bork P (2008) Predicting biological networks from genomic data. FEBS Lett 582(8):1251–1258

Harris MA, Clark J, Ireland A, Lomax J, Ashburner M, Foulger R, Eilbeck K, Lewis S, Marshall B, Mungall C, Richter J, Rubin GM, Blake JA, Bult C, Dolan M, Drabkin H, Eppig JT, Hill DP, Ni L, Ringwald M, Balakrishnan R, Cherry JM, Christie KR, Costanzo MC, Dwight SS, Engel S, Fisk DG, Hirschman JE, Hong EL, Nash RS, Sethuraman A, Theesfeld CL, Botstein D, Dolinski K, Feierbach B, Berardini T, Mundodi S, Rhee SY, Apweiler R, Barrell D, Camon E, Dimmer E, Lee V, Chisholm R, Gaudet P, Kibbe W, Kishore R, Schwarz EM, Sternberg P, Gwinn M, Hannick L, Wortman J, Berriman M, Wood V, de la CN, Tonellato P, Jaiswal P, Seigfried T, White R (2004) The Gene Ontology (GO) database and informatics resource. Nucleic Acids Res 32(Database issue):D258–D261

Hayat S, Elofsson A (2012) BOCTOPUS: improved topology prediction of transmembrane β barrel proteins. Bioinformatics 28(4):516–522

Henikoff S, Henikoff JG (1992) Amino acid substitution matrices from protein blocks. Proc Natl Acad Sci U S A 89:10915–10919

Holm L, Rosenström P (2010) Dali server: conservation mapping in 3D. Nucleic Acids Res 38:W545–W549

Holm L, Kaariainen S, Wilton C, Plewczynski D (2006) Using Dali for structural comparison of proteins. Curr Protoc Bioinformatics Chap. 5:Unit 5.5

Huang DW, Sharman BT, Lempicki RA (2009) Bioinformatics enrichment tools: paths toward the comprehensive functional analysis of large gene lists. Nucleic Acids Res 37(1):1–13

Hunter S, Jones P, Mitchell A, Apweiler R, Attwood TK, Bateman A, Bernard T, Binns D, Bork P, Burge S, de Castro E, Coggill P, Corbett M, Das U, Daugherty L, Duquenne L, Finn RD, Fraser M, Gough J, Haft D, Hulo N, Kahn D, Kelly E, Letunic I, Lonsdale D, Lopez R, Madera M, Maslen J, McAnulla C, McDowall J, McMenamin C, Mi H, Mutowo-Muellenet P, Mulder N, Natale D, Orengo C, Pesseat S, Punta M, Quinn AF, Rivoire C, Sangrador-Vegas A, Selengut

JD, Sigrist CJ, Scheremetjew M, Tate J, Thimmajanarthanan M, Thomas PD, Wu CH, Yeats C, Yong SY (2012) InterPro in 2011: new developments in the family and domain prediction database. Nucleic Acids Res 40(Database issue):D306–D312. doi:10.1093/nar/gkr948

IUPAC-IUB (1969) Commission on Biochemical Nomenclature. A one-letter notation for amino acid sequences. Tentative rules. Biochem J 113 (1):1–4

Jones DT (1999) Protein secondary structure prediction based on position-specific scoring matrices. J Mol Biol 292:195–202

Käll L, Krogh A, Sonnhammer ELL (2007) Advantages of combined transmembrane topology and signal peptide prediction-the Phobius web server. Nucleic Acids Res 25:W429–W432

Kanehisa M, Goto S, Kawashima S, Okuno Y, Hattori M (2004) The KEGG resource for deciphering the genome. Nucleic Acids Res 32(Database issue):D277–D280

Katoh K, Standley DM (2013) MAFFT multiple sequence alignment software version 7: improvements in performance and usability. Mol Biol Evol 30(4):772–780

Krissinel E, Henrick K (2004) Common subgraph isomorphism detection by backtracking search. Softwa Pract Exp 34(6):591–607

Krissinel E, Henrick K (2005) Multiple alignment of protein structures in three dimensions. In: Berthold MR et al (eds) CompLife 2005, LNBI 3695. Springer, Berlin, pp 67–78

Krogh A, Larsson B, Heijne Gv, Sonnhammer ELL (2001) Predicting transmembrane protein topology with a hidden Markov model: application to complete genomes. J Mol Biol 305(3):567–580

Laskowski RA, Rullmannn JA, MacArthur MW, Kaptein R, Thornton JM (1996) AQUA and PROCHECK-NMR: programs for checking the quality of protein structures solved by NMR. J Biomol NMR 8(4):477–486

Laskowski RA, Chistyakov VV, Thornton JM (2005) PDBsum more: new summaries and analyses of the known 3D structures of proteins and nucleic acids. Nucleic Acids Res 33:D266–D268

Letunic I, Bork P (2006) Interactive Tree Of Live (iTOL): an online tool for phylogenetic tree display and annotation. Bioinformatics 32(1):127–128

Lupas A, Dyke Mv, Stock J (1991) Predicting coiled coils from protein sequences. Science 252:1162–1164

McGuffin LJ, Buenavista MT, Roche DB (2013) The ModFOLD4 server for the quality assessment of 3D protein models. Nucleic Acids Res 41(W1):W368–W372

Najmanovich R, Kurbatova N, Thornton J (2008) Detection of 3D atomic similarities and their use in the discrimination of small-molecule protein binding sites. Bioinformatics 24:105–111

Needleman SB, Wunsch CD (1970) A general method applicable to the search for similarities in the amino acid sequence of two proteins. J Mol Biol 48:443–453

Neumann S, Fuchs A, Mulkidjanian A, Frishman D (2010) Current status of membrane protein structure classification. Proteins 78(7):1760–1773

Notredame C, Higgins DG, Heringa J (2000) T-Coffee: a novel method for fast and accurate multiple sequence alignment. J Mol Biol 302(1):205–217

Papadopoulos JS, Agarwala R (2007) COBALT: constraint-based alignment tool for multiple protein sequences. Bioinformatics 23(9):1073–1079. doi:10.1093/bioinformatics/btm076

Pearl F, Todd A, Sillitoe I, Dibley M, Redfern O, Lewis T, Bennett C, Marsden R, Grant A, Lee D, Akpor A, Maibaum M, Harrison A, Dallman T, Reeves G, Diboun I, Addou S, Lise S, Johnston C, Sillero A, Thornton J, Orengo C (2005) The CATH domain structure database and related resources Gene3D and DHS provide comprehensive domain family information for genome analysis. Nucleic Acids Res 33(Database issue):D247–D251

Petersen TN, Brunak S, Heijne Gv, Nielsen H (2011) SignalP 4.0: discriminating signal peptides from transmembrane regions. Nat Methods 8:785–786

Pietrosemoli N, Lopez D, Segura-Cabrera A, Pazos F (2012) Computational prediction of important regions in protein sequences. IEEE Signal Process Mag 29(6):143–147

Prlic A, Bliven S, Rose PW, Bluhm WF, Bizon C, Godzik A, Bourne PE (2010) Precalculated protein structure alignments at the RCSB PDB website. Bioinformatics 26(23):2983–2985

Rice P, Longden I, Bleasby A (2000) EMBOSS: the European molecular biology open software suite. Trends Genet 16(6):276–277

Roy A, Kucukural A, Zhang Y (2010) I-TASSER: a unified platform for automated protein structure and function prediction. Nat Protoc 5:725–738

Sanchez R, Serra F, Tarraga J, Medina I, Carbonell J, Pulido L, de Maria A, Capella-Gutierrez S, Huerta-Cepas J, Gabaldon T, Dopazo J, Dopazo H (2011) Phylemon 2.0: a suite of web-tools for molecular evolution, phylogenetics, phylogenomics and hypotheses testing. Nucleic Acids Res 39(Web Server issue):W470–474

Sander C, Schneider R (1991) Database of homology-derived structures and the structural meaning of sequence alignment. Proteins 9:56–68

Schneider TD, Stephens RM (1990) Sequence logos: a new way to display consensus sequences. Nucleic Acids Res 18(20):6097–6100

Schomburg I, Chang A, Ebeling C, Gremse M, Heldt C, Huhn G, Schomburg D (2004) BRENDA, the enzyme database: updates and major new developments. Nucleic Acids Res 32(1):D431–D433

Schwede T, Kopp J, Guex N, Peitsch MC (2003) SWISS-MODEL: an automated protein homology-modeling server. Nucleic Acids Res 31:3381–3385

Shindyalov IN, Bourne PE (1998) Protein structure alignment by incremental combinatorial extension (CE) of the optimal path. Protein Eng 11(9):739–747

Sievers F, Wilm A, Dineen D, Gibson TJ, Karplus K, Li W, Lopez R, McWilliam H, Remmert M, Soding J, Thompson JD, Higgins DG (2011) Fast, scalable generation of high-quality protein multiple sequence alignments using Clustal Omega. Mol Syst Biol 7:539. doi:10.1038/msb.2011.75

Sigrist CJ, Cerutti L, Hulo N, Gattiker A, Falquet L, Pagni M, Bairoch A, Bucher P (2002) PROSITE: a documented database using patterns and profiles as motif descriptors. Brief Bioinform 3(3):265–274

Smith TF, Waterman MS (1981) Identification of common molecular subsequences. J Mol Biol 147:195–197

Sobolev V, Sorokine A, Prilusky J, Abola EE, Edelman M (1999) Automated analysis of interatomic contacts in proteins. Bioinformatics 15:327–332

Soding J, Biegert A, Lupas AN (2005) The HHpred interactive server for protein homology detection and structure prediction. Nucleic Acids Res 33(Web Server issue):W244–248. doi:10.1093/nar/gki408

UniProt C (2014) Activities at the Universal Protein Resource (UniProt). Nucleic Acids Res 42(Database issue):D191–D198. doi:10.1093/nar/gkt1140

Vicent TL, Green PJ, Woolfson DN (2013) LOGICOIL-multi-state prediction of coiled-coil oligomeric state. Bioinformatics 29(1):69–76

von Heijne G (2006) Membrane-protein topology. Nat Rev Mol Cell Biol 7(12):909–918

von Mering C, Huynen M, Jaeggi D, Schmidt S, Bork P, Snel B (2003) STRING: a database of predicted functional associations between proteins. Nucleic Acids Res 31:258–261

Wallace AC, Laskowski RA, Thornton JM (1995) LIGPLOT: a program to generate schematic diagrams of protein-ligand interactions. Protein Eng 8(2):127–134

Ward JJ, Sodhi JS, McGuffin LJ, Buxton BF, Jones DT (2004) Prediction and functional analysis of native disorder in proteins from the three kingdoms of life. J Mol Biol 337(3):635–645

Waterhouse AM, Procter JB, Martin DM, Clamp M, Barton GJ (2009) Jalview Version 2-a multiple sequence alignment editor and analysis workbench. Bioinformatics 25(9):1189–1191

Yamada T, Letunic I, Okuda S, Kanehisa M, Bork P (2011) iPath2.0: interactive pathway explorer. Nucleic Acids Res 39:W412–W415

Ye Y, Godzik A (2003) Flexible structure alignment by chaining aligned fragment pairs allowing twists. Bioinformatics 9(S2):ii246–ii255

Index

F. Pazos, M. Chagoyen, *Practical Protein Bioinformatics*,
DOI 10.1007/978-3-319-12727-9